Materials Enabled Designs

Materials Enabled Designs

The Materials Engineering Perspective to Product Design and Manufacturing

Michael Pfeifer

AMSTERDAM • BOSTON • HEIDELBERG • LONDON
NEW YORK • OXFORD • PARIS • SAN DIEGO
SAN FRANCISCO • SINGAPORE • SYDNEY • TOKYO

Butterworth-Heinemann is an imprint of Elsevier

Butterworth-Heinemann is an imprint of Elsevier
30 Corporate Drive, Suite 400
Burlington, MA 01803

This book is printed on acid-free paper. ∞

Library of Congress Cataloging-in-Publication Data
Application submitted.

ISBN 13: 978-0-7506-8287-9

For information on all Butterworth-Heinemann publications,
visit our Website at *www.books.elsevier.com*

Printed in the United States
09 10 11 12 13   10 9 8 7 6 5 4 3 2 1

To Jenny, Jordan, and David

Contents

Preface

This book covers the materials considerations required to improve the likelihood of designing, developing, and manufacturing successful products. Such considerations are important in view of the significant role that materials play in the success of a product and the many decisions during product development and manufacturing that influence the performance, reliability, and cost of the materials used in a product. Some of these decisions include product design concept selection, materials selection, manufacturing process selection, and supplier selection.

The idea for this book came about after I taught a class in the Manufacturing and Design Engineering (MaDE) program at Northwestern University. The course focused on the materials engineering considerations for product design, development, and manufacturing. As I assembled the reading material for the class, I found that there were no texts that addressed product development and manufacturing from the materials engineering perspective.

There are several books about product design, but they are written from the mechanical engineering perspective. While some of these books discuss materials selection, they do so from a mechanical engineering viewpoint. That is, they discuss the process for selecting materials based on satisfying product performance requirements, but they neglect the many other design requirements that must be considered when selecting materials.

Other books discuss materials selection, but they do not cover all of the applicable design requirements and do not discuss the process of verifying that the materials do indeed satisfy all of the design requirements. Also, these books do not address in detail the materials engineering considerations for developing capable manufacturing processes and evaluating the reliability of materials for specific designs.

The concepts presented here complement the information provided in product design and materials selection textbooks. This book also complements books that focus on other design considerations such as design for manufacturing, design for reliability, and design for environmental variables. The only difference is that this book focuses on the materials aspects of the design for X approaches.

To avoid confusion and manage reader expectations, it is important to mention what is and is not presented here. First, this book's focus is on the materials engineering considerations for specific decisions made during product development and manufacturing; that is, only the decisions that benefit from the materials engineering perspective are considered. Second, the process and considerations for materials selection are covered; however, the selection of materials for specific applications is not covered because plenty of books are available on that topic.

Chapter 1 explains the materials engineering perspective; the role of materials and materials engineering in a product; and how a product is ultimately an assemblage of materials that must be selected and whose properties must be controlled. The chapter also defines terms used throughout the book.

Chapter 2 discusses the design requirements that the materials in a product must satisfy and explains how the requirements are derived from the wants and needs of the product's intended customer. Chapter 3 outlines the process of choosing materials based on materials selection criteria.

Chapters 4 through 6 present background information about materials engineering and related considerations for performance, reliability, and product manufacturing. Chapter 4 discusses the aspects of materials that must be controlled to obtain the desired properties and the resources available for technical information about materials.

Chapter 5 covers the aspects of manufacturing processes that influence the properties, performance, and reliability of the materials that go through a manufacturing process. This chapter briefly discusses various manufacturing processes, explores the general aspects of manufacturing processes that must be controlled in order for the materials that make up a product to be as desired, and addresses manufacturing process variations and their impact on the materials that constitute the process output. Chapter 6 examines the reliability of materials and presents strategies for evaluating that reliability.

Chapters 7 through 12 apply the information provided in the previous chapters to the various elements of product development and manufacturing that require the materials engineering perspective.

Acknowledgments

I would like to start by thanking Professor Ed Colgate from Northwestern University for his support and encouragement. Ed took me up on my idea to offer a course based on the materials engineering considerations for product development for the Manufacturing and Design Engineering (MaDE) program. This book is based on the material from that course. I also want to thank Ed for his insightful review of it. His comments and suggestions resulted in dramatic improvements.

Next, I want to thank Ron Scicluna. His insights and knowledge about the product development process were critical to helping me better understand the place of materials engineering and the importance of risk-assessment and mitigation strategies throughout every phase of the process. The many hours of discussion with Ron were educational and fun, and they helped me organize my thoughts.

I want to thank Stan Rak, Steve Gonczy, and Dmitriy Shmagin for reviewing various portions of this book. All of their comments and suggestions were useful. Steve also helped me prepare the section on ceramics in Chapter 4. Craig Miller and Stacey Mosley, both students in the MaDE class at Northwestern, also provided valuable feedback.

Many of those who helped me obtain some of the images used here went above and beyond their duty to provide assistance. These people are Scott Henry and Ann Britton from ASM International, Michael Sagan and Michael Hammond from Trek Bikes, Ed Wolfe from ANH Refractories, Anita Brown from Indium Corporation, and Tim Dyer from Carpenter Advanced Ceramics. Also, David Zukerman used his vast graphic arts skills to help me get some of the images ready for production.

Marilyn Rash was the project manager at Elsevier for this book. Her editing made many concepts clearer and reduced the redundancies that I liberally scattered throughout the book. Also, although I was late in getting reviewed portions of the book back to her, Marilyn was still able to keep the book production on track.

Finally, to my wife, Jenny, thank you for supporting my efforts to write this book. It seemed at times like the writing would never end.

The Materials Engineering Perspective

1.1 INTRODUCTION

A person can look at any engineered product and see that it is made of a wide variety of materials that have been manipulated into a wide variety of shapes for the purpose of enabling specific product features. Just consider an automobile with its painted steel body, plastic knobs, rubber tires, and glass windows, or a computer mouse with its plastic shell and buttons and rubber tracking ball and wheel, or a bicycle with its painted aluminum frame, steel gears and chain, and foam padded and plastic covered seat. In fact, a product can be considered to be a collection of materials such as metals, polymers, ceramics, composites, and semiconductors. Furthermore, the materials used in a product account for up to 60% of the total cost to manufacture a product (Nevins & Whitney, 1989). Based on both of these facts, it seems that the engineering processes for selecting the materials used in a product and the means by which the properties of the materials are controlled are of the utmost importance to the success of a product.

Even though the materials used in a product have a huge impact on its performance, reliability, and cost, many companies vastly undervalue the importance of proper materials engineering considerations for product development and manufacturing decisions. Consequently, these companies struggle with problems such as new products that are behind schedule, cost overruns, poor supplier quality, poor manufacturing quality, and products that do not work as expected. All of these problems have a negative effect on the success of a product and a company's competitiveness. These struggles do not have to be accepted as a normal part of doing business. In many cases, product development and manufacturing problems, and their costs, can be avoided if comprehensive materials engineering considerations are employed when making certain design and manufacturing decisions.

A successful product enjoys good profits, good market share, and good customer satisfaction. Developing and manufacturing a successful product requires the following:

- That the product has the performance and reliability to satisfy the wants and needs of the intended customer
- That the costs to develop and manufacture the product are within budget
- That the product is released to the market on time

Meeting the first two requirements depends on a design team's ability to select materials that enable the product to satisfy its performance, reliability, and cost requirements. Furthermore, controlling the variation of the properties of the materials is critical for making a product that consistently meets its performance and reliability requirements while keeping manufacturing costs within budget. Releasing a product to market on time depends on avoiding delays associated with problems with the materials.

In short, this book asserts that a product's success depends on the attention paid to the materials engineering aspects of decisions that occur during product development and manufacturing.

It is not the intention here to diminish the role of other engineering perspectives or to imply that materials engineering alone can solve all the problems encountered during product development and manufacturing. The materials engineering perspective is just one perspective of many that are required to make good decisions that increase the likelihood of producing a successful product. However, it is the intention of this book to instill a better appreciation for the role that the materials engineering perspective can play in product success.

1.2 THE MATERIALS ENGINEERING PERSPECTIVE

This book teaches a perspective that focuses on materials engineering concerns as they pertain to achieving overall product success. This perspective, referred to here as the *materials engineering perspective,* is based on the following three considerations:

1. The *performance, reliability,* and *cost* of a product are highly dependent on the properties of the materials that make up the product.
2. Proper *selection* of the materials used in a product is crucial to satisfy its performance, reliability, and cost requirements.
3. *Control of the variation* of the properties of the materials that make up a product is crucial for enabling its consistent performance, reliability, and cost.

The first consideration is important because it shifts the attention away from viewing any single component within a product solely in terms of its mechanical, electrical, optical, or chemical functionality. Instead, seeing a component in terms of its materials moves attention to the properties of the materials required to obtain the desired functionality and reliability at the desired cost.

The second consideration may seem obvious because most engineers recognize that specific materials have specific applications and that the optimum materials must be selected for any given application. However, the proper selection of materials demands thorough and accurate knowledge of all of a product's performance, reliability, and cost requirements. Many design teams make the mistake of trying to select materials without knowing all the selection criteria and based on inaccurate criteria. Furthermore, there are selection criteria that are based on requirements in addition to performance, cost, and reliability. For example, industry standards, government regulations, intellectual property rights, and manufacturing constraints place requirements on a product's design. This is discussed in more detail in Chapter 2.

The third consideration about the control of material properties is based on the fact that there are many sources of variation of the properties of the materials used in a product. The sources of variation are related to the processes used to manufacture a product and the materials used in the processes. Controlling variations requires an understanding of the relationship between a manufacturing process, the properties of materials used in the process, and the properties of the material that makes up the process output. Excessive variations in the materials' properties result in products that cannot be easily manufactured and do not have the desired performance and reliability. This is discussed in more detail in Chapter 5.

Looking at a product from the materials engineering perspective can help design teams frame decisions and understand the information required to make better design and manufacturing decisions. An example of the application of this perspective can be provided through consideration of the scissors shown in Figure 1.1. From just a functional perspective, the scissors is a mechanical device capable of cutting paper. From a materials engineering perspective, the scissors is a set of materials that must have certain properties, such as the following.

FIGURE 1.1

Pair of scissors.

- Two pieces of corrosion-resistant material hard enough to maintain a sharp edge and ductile enough so as not to fracture when used to pry something open
- Handles rigid enough to transfer a user's force to the blades, but with enough strength and impact resistance so that they do not crack or break when the scissors are used or dropped
- A pivot pin made of a hard, corrosion-resistant material with a surface smooth enough so that the blades pivot with little effort

Furthermore, there are common requirements for all the materials. Namely, that the materials enable the blades, handles, and pivot pin to be easily manufactured and that the materials are of reasonable cost.

Recognition of all these requirements and their importance helps engineering teams focus on the possible materials that can be considered for use and selecting the materials that optimize a product's performance, reliability, and cost to produce.

The materials engineering perspective also helps engineering teams focus on how to control the variation of the material properties to ensure that a product consistently satisfies the wants and needs of the customer. This involves understanding the effects of variations in the manufacturing process on the materials' properties variations, developing capable manufacturing processes, and selecting capable suppliers.

Now, imagine designing more complicated products that have performance and reliability requirements that are much more demanding than for a pair of scissors (e.g., a jet engine, a hip implant, or an automobile fuel level sensor) and that are exposed to much harsher environments. What is the likelihood of the success of these products if the optimum materials are not selected and are not well controlled?

The materials engineering perspective may seem like a narrow topic on which to write a book aimed at product design, development, and manufacturing. However, many decisions occur during product design, development, and manufacturing that have an impact on the materials selected for use in a product and how well the properties of the materials are controlled. These decisions will be discussed from the materials engineering perspective. The chances of these decisions resulting in favorable outcomes improves when a materials engineering perspective is brought into the decision-making process.

This book is different from others on materials engineering in that the science and engineering of materials is not the focus. Instead, the focus here is on the considerations and information required to make better and faster decisions that affect the materials used in a product. These decisions occur throughout every phase of product design, development, and manufacturing. Furthermore, these decisions go well beyond just material selection and failure analysis—two aspects of the product life cycle that are associated with materials engineering. Some of the decisions that will benefit from a materials engineering perspective will seem obvious. Others are not as obvious and may even appear counterintuitive at first. However, the discussion of the materials engineering perspective for the specific decision will illustrate its significance. Only those decisions that involve or impact the materials are considered here.

The information in this book conveys how knowledge of materials engineering and the materials engineering perspective can provide a competitive advantage that will reduce the costs and time to develop and manufacture a product. However, readers should be aware of the subjects on which this book does not focus.

First, even though certain aspects of the design process are discussed, what is here will not teach product design and development. Instead, it is intended as a complement to textbooks that focus on product design and development (e.g., Ullman, 2003; Ulrich & Eppinger, 2004; Pahl & Beitz, 1996). Second, although we provide some explanation of materials science and materials properties, it is not the purpose of this book to teach materials science or materials selection for specific applications. Resources for this information will be provided in later chapters.

The concepts discussed here are in practice at a few companies. At those companies, new products are brought to market with fewer problems compared to companies that do not have materials engineers. Also, new materials for performance improvement, reliability improvement, and cost reduction are continually being evaluated and implemented.

1.3 WHAT IS MATERIALS ENGINEERING?

In order to understand the materials engineering perspective it is helpful to understand what materials engineering is. It involves understanding the relationship between the properties of a material, its composition, its microscopic structures, and how it was processed. This knowledge is put to use to develop and improve products and manufacturing processes.

Materials engineering education includes the study of the following:

- Microscopic structures within materials
- Atomic and molecular motion, and the interactions and reactions between atoms and molecules within materials
- Macroscopic material properties
- Effects of microscopic structures and composition on macroscopic material properties
- Effects of the manufacturing processes on the microscopic structure and macroscopic properties
- Methods for characterizing microscopic structures and macroscopic properties
- Mechanisms by which materials degrade

The macroscopic properties of a material depend on its microscopic structure and on its composition, which is discussed in more detail in Chapter 4. Degradation mechanisms of materials, which include corrosion, mechanical yielding, fatigue, wear, and electrical breakdown, is discussed in more detail in Chapter 6.

Materials engineers work with engineers from other disciplines (e.g., mechanical, electrical, biomedical, and aerospace) to understand the performance and reliability requirements of a product. With this information, a materials engineer can help identify a material or set of materials that might meet the performance and reliability requirements, in addition to cost, manufacturing, compliance, and availability requirements. Narrowing the list of potential materials to those known to meet all the requirements is the next step, which is discussed in the following chapters.

Materials engineers apply their knowledge and experience toward many aspects of product design, development, and manufacturing. They ply their trade as design, product reliability, manufacturing, manufacturing and supplier quality, and failure analysis engineers, and in some cases as purchasing agents.

In spite of the many decisions that could benefit from the materials engineering perspective and expertise, the problems that can be prevented, and the increased profits that can be earned, many organizations overlook the need to gain a solid understanding of the science and engineering of the materials used in their products. Instead, those with backgrounds in other engineering disciplines are called on to act as materials engineers to make materials engineering decisions. The result is that many relevant and necessary materials issues are simply not considered during product development and manufacturing. Consequently, many products suffer from the problems discussed in the previous section.

Nonmaterials engineers give several reasons for not seeking out materials engineering expertise. All of them are common in that they are based on certain myths about the perceived need for materials engineering and the experience and perspective required for making good decisions where the selection and control of materials are concerned. These myths are as follows:

Myth 1. Materials engineering only involves lab and failure analyses. A result of this mind-set is that companies that do not have materials engineers on staff only call on one for assistance when there is a problem, rather than proactively seeking help during the design phase of a product, during process development, or when evaluating a new supplier.

Myth 2. Materials engineering considerations are only needed for "high-tech" materials and applications. The labs of materials testing services are filled with failed products that do not meet design requirements, many of which are composed of common materials (e.g., steel, aluminum alloys, polypropylene, nylon, epoxy, silicon, nickel plating, and paint). These examples constitute a very small portion of the huge list of "low-tech" materials that are incorrectly selected, specified, and manufactured. The reality is, materials engineering is applicable and useful even for common materials. In some cases, the considerations are more complex than others and depend on a material's performance requirements, the manufacturing processes used to shape a material into its final form, how it is used, and the conditions to which it is exposed during use.

Myth 3. Materials engineering decisions are intuitive and anyone can make these decisions. The countless number of product recalls, field failures, missed product deadlines, poor supplier quality, and manufacturing quality issues indicates that intuition is not sufficient for making sound materials engineering decisions. Engineers should keep in mind that materials engineering is a discipline for which people earn B.S., M.S., and Ph.D. degrees and then continue to apply the learning on a daily basis. In contrast, engineers in other disciplines might have taken one or two materials science classes in college and occasionally apply this learning.

Myth 4. Suppliers can be relied on for materials engineering support. This may be true if suppliers have materials engineers on staff and if they can take the time to address the decision or problem being considered. However, many suppliers do not have materials engineers or they are focused on developing their companies' products. Although a supplier without materials engineers on staff may be able to provide some guidance based on past experiences for general applications, they often do not have enough available expertise to address applications that extend a short distance beyond previous experience. When some on-staff person is able to provide technical support, it is based on rote experience, not on fundamental materials science knowledge.

As discussed in the introduction, decisions will have a higher probability of success if all the necessary perspectives are represented by people with the appropriate backgrounds and experiences. The costs in terms of time and money to do this are much less compared to the long-term costs of a bad decision.

1.4 PRODUCTS AND THEIR MATERIALS

A major premise here is that an appreciation of the materials engineering perspective will help engineers better understand the risks and rewards associated with an "informed selection" of the materials used in a product; the manufacturing processes used to produce the product; and the suppliers of materials, components, and subassemblies used in a product. This "understanding" will help define relevant risk-mitigation strategies and manage expectations.

The ability to recognize and internalize that a product is an assemblage of materials that can enable the product's success is one key to understanding the materials engineering perspective. Figure 1.2 shows how an assembly can be broken down into subassemblies, components, joints, in-process structures, and materials. An *assembly* is defined here as a complete technical system that is used

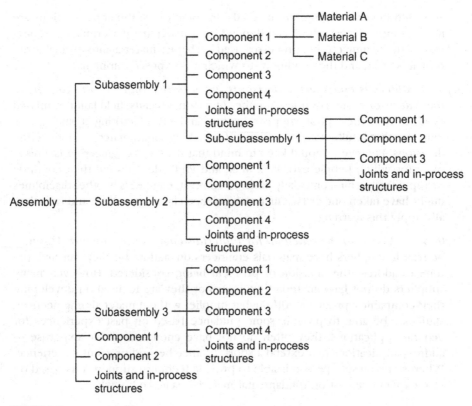

FIGURE 1.2

Diagram for defining assembly, subassembly, and component.

to perform one or more tasks and is composed of subassemblies or components. Examples of assemblies are automobiles, airplanes, manufacturing equipment, garden tools, computers, home appliances, furniture, and consumer electronics. Computer monitors and other computer peripheral equipment, stereo speakers, and automobile wheels are not considered to be assemblies because by themselves they cannot be used to perform any tasks.

A *subassembly* is an assembled set of components that may provide some of the functionality of the assembly. A subassembly is assembled before its incorporation into an assembly. Subassemblies can be composed of components that have been joined together, other subassemblies that have been joined together, or components and other subassemblies that have been joined together. For very complex assemblies, such as an airplane or automobile, there are several levels of subassemblies (sub-subassemblies and so on). Thus, computer monitors, stereo speakers, and automobile wheels are subassemblies; so are engines for lawnmowers and motorcycles, pumps, motors for electric drills and fans, lightbulbs, and electrical circuit boards.

A *component* is defined as a singular fabricated item that does not contain other fabricated items as part of its structure. It does not consist of two or more items that have been joined together. A component is composed of one or more materials, such as metal, ceramic, or polymer, and it can be fabricated using one or more manufacturing processes.

Components and subassemblies are joined together by mechanical and non-mechanical methods. Mechanical joints can involve the use of an additional component in the form of a fastener such as a screw or rivet. There are also mechanical joining methods, such as snapping or crimping components together, that do not involve an additional component. Examples of both types of mechanical joints are shown in Figure 1.3.

Nonmechanical joints are formed using methods that employ adhesion, metallurgical reactions, or chemical reactions between the components or subassemblies being joined. The methods for forming these types of joints include the following:

- Adhesive bonding using adhesives such as tape, epoxy, or silicone
- Soldering or brazing using solders or braze compounds, respectively
- Welding with or without weld filler material
- Diffusion bonding

Figure 1.4 shows schematics of the different nonmechanical joining methods. In contrast to the mechanical methods of joining, they involve bonding mechanisms that occur at the interface between the bonding material and each of the

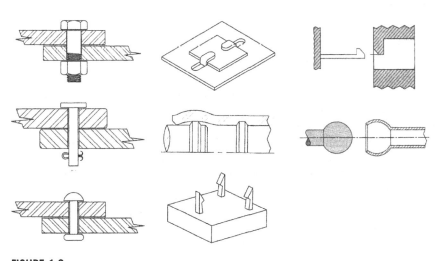

FIGURE 1.3

Examples of mechanical joints.

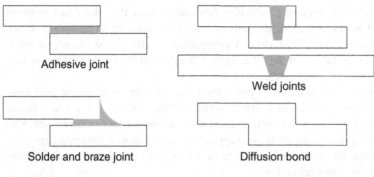

FIGURE 1.4

Schematics of nonmechanical joints.

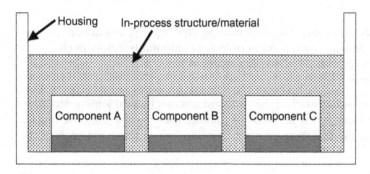

FIGURE 1.5

Schematic of an in-process structure.

components being joined or that involve reactions or interactions between the components being joined.

During the process of making a subassembly or assembly, materials can be applied onto one or more components or subassemblies for the purpose of coating or forming some other structure on the subassembly or assembly. The material applied to the subassembly or assembly will be referred to as an *in-process material*, and the coating or structure formed will be referred to as an *in-process structure*. A schematic of an in-process structure is shown in Figure 1.5; the following are some examples:

- Silicone coating used to coat electronic circuit boards after they have been populated with electrical components. The coating prevents moisture from getting onto the components underneath.

- Encapsulants that are used to fill a subassembly cavity for the purpose of preventing moisture and chemicals from getting inside the subassembly.

■ Varnish added to electrical laminations to provide electrical insulation.

■ Thermally conductive material injected between two components to aid in heat removal during product operation.

So, there are three basic elements that make up a product: components, joints, and in-process structures. The materials and methods used to produce these *product elements* must be selected so that the design of each product element is optimized, enabling the proper performance of the product in which the product element is used. Furthermore, the materials and methods used to produce a product element must be controlled so that it always meets its design requirements, enabling the manufacture of a product that always meets its design requirements.

An example of an assembly that consists of various levels of subassemblies is a bicycle (Figure 1.6). A bicycle has frame, wheel and tire, crankset, gear shift levers, front derailleur, rear derailleur, saddle, chain, brake, and brake lever subassemblies. Figure 1.7 shows an exploded view of the bicycle crankset, demonstrating the various components within its subassembly. An example of a bicycle subassembly that contains other subassemblies is the wheel and tire subassembly, which consists of a rim, rim strip, spoke, spoke nipple, tire, inner tube, and hub subassembly.

Some products are complicated and have many subassemblies (e.g., airplanes and automobiles). Other less complicated products, such as a power drill or a radio, have fewer subassemblies. And others are simple products, such as a lead pencil, a screwdriver, or a pair of scissors. Regardless of a product's complexity, its success depends on selecting materials that enable the product to consistently meet its performance, reliability, and cost requirements.

In this book the word *product* is defined as an item that one company sells to an end user or sells to another company for incorporation into the purchasing

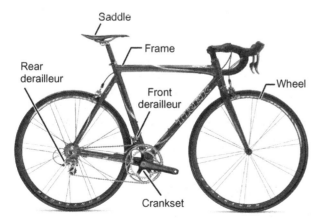

FIGURE 1.6

Subassemblies of a bicycle. (*Source:* Reprinted with the permission of Trek Bikes.)

FIGURE 1.7

Exploded view of the crankset for the bicycle shown in Figure 1.6. (*Source:* Reprinted with the permission of Trek Bikes, Michael Hammond designer.)

company's product as a subassembly or component. A product can also be a material that is used to fabricate components or join components and subassemblies or form an in-process structure. Products sold to end users include consumer electronics, appliances, hand tools, factory equipment, home and office furniture, paint, lightbulbs, and soap. Products sold for use in another product include screws, cast components, motors, electrical components, valves, bearings, pressure gauges, machined parts, and engineering materials.

Sometimes the word "product" is too ambiguous. When this is the case, the words "assembly," "subassembly," and "component" will be used. However, even these more specific words mean different things to different people. For example, whereas one company may view its product as an *assembly*, the customer may view it as a *component* or *subassembly* to be used within its own product. So a company that makes computer cables may view its product as being an *assembly*, whereas a computer manufacturer may view a computer cable as being a *component*.

1.5 PRODUCT SUCCESS AND THE MATERIALS ENGINEERING PERSPECTIVE

The materials engineering perspective should be of significance to business leaders and engineers who work at a wide range of manufacturing companies and work on different types of product development projects. Implementing this perspective will help design teams identify risks to the success of a product, develop strategies to mitigate the risks, and identify opportunities to optimize the design.

An organization's ability to develop and manufacture successful products depends on the ability of its business and engineering leaders to consistently make good business and engineering decisions regarding products to develop, their design, and how to manufacture them. A significant number of these decisions influence whether or not the materials used in a product satisfy the design requirements and the variation of the properties of the materials. Some of these decisions and their impact on the materials used are as follows:

Selecting the mechanical and electrical design. This influences the options of materials that can be considered for use.

Selecting the materials. This influences (1) the performance, reliability, and cost of each product element; (2) the variations of the performance and reliability of each product element; and (3) the ease of manufacturing product elements and subassemblies.

Selecting manufacturing processes. This influences the options of materials that can be considered for use.

Developing manufacturing processes. This influences the variation of the material properties for product elements.

Selecting suppliers of components and subassemblies. This influences the material properties of product elements and the variation of the properties.

These decisions and the materials engineering perspective for the decisions are discussed in detail in Chapters 7 through 11.

For business leaders, success is ultimately measured by the financial return on the investments required to develop and manufacture a product. Obtaining a good return on investment requires the following critical success factors:

- Selecting opportunities that best suit an organization's abilities
- Meeting product development schedules
- Meeting product cost targets
- Generating products that have good customer acceptance and market share

For engineering teams, success is measured by the ability to consistently develop products that meet the following criteria:

- Satisfy their performance, reliability, and cost requirements
- Satisfy all the customer's wants and needs
- Have good manufacturing yields

Meeting the goals of the business leaders and engineering teams requires making good decisions during the entire design process. They must be able to identify risks to the success of a product, develop strategies to mitigate the risks, and determine whether any of the risks are unacceptable.

During the design process it is easier, and less costly, to make design changes early in the process rather than later. However, at the beginning of the project the knowledge is not yet available to make optimum decisions. As time passes, a design team gains knowledge; however, it is more difficult, and expensive, to make changes to a product's design because commitments have been made regarding manufacturing processes, equipment, and the tooling and suppliers to use. Any changes in these commitments will result in delays to the product development effort and potentially extra costs.

The relationship between ease of change and knowledge is called the *design paradox* (Ullman, 2003), which is illustrated in Figure 1.8; it shows the ease of making design changes and knowledge gained as a function of time into the design process. To be competitive, design teams must look for ways to gain the required knowledge early in the design process, enabling faster, better-informed decisions.

Excessive lack of knowledge in the early phases of product development leads to the risks to success that are encountered. It is important to have knowledge of the risks as early as possible and to develop strategies to mitigate them. As a project progresses, more resources are added to deal with the increasing number of design decisions that the team must make. At the same time, newer risks will be identified, but the total number of risks should decrease as the project team gains more knowledge about the product's design.

The relationship between risks, resources invested, and time into the design process is shown in Figure 1.9. If at any point it is determined that the risks are

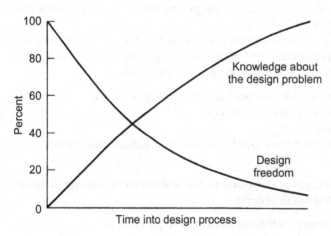

FIGURE 1.8

Design paradox. (*Source:* Adapted from Ullman, 2003; reprinted with the permission of McGraw-Hill.)

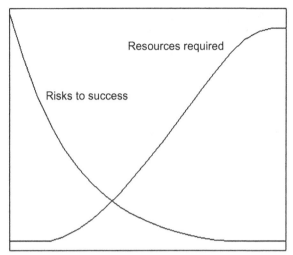

FIGURE 1.9

Risks and resources.

too high to meet the business and engineering goals, then either the project should be abandoned and the investment made elsewhere, or the business opportunity must be redefined. More resources should be invested only if it is possible to mitigate the risks.

Including the materials engineering perspective when making decisions is an effective strategy to obtain required knowledge and identify risks early in a project. This will lead to engineering trade-offs and compromises during engineering deliberations, which will improve the chances of making decisions that will satisfy all the engineering perspectives and optimize the product's design. This applies to all of the perspectives that impact the success of a product, such as the viewpoints of representatives from marketing, supply chain, product reliability, and manufacturing.

To illustrate this point, consider the Venn diagram shown in Figure 1.10. Each circle represents the possible choices offered by different perspectives for a decision. For design and manufacturing decisions, possible perspectives include those from mechanical, electrical, materials, and manufacturing engineering, as well as those from marketing, quality assurance, and purchasing. The region where all three circles overlap (*gray area*) corresponds to the set of choices that provide the optimum outcome. If only two of the three perspectives are considered, then selecting any of the suboptimum choices (*dotted area*) becomes a possibility.

Consider a product design example for an electromechanical device whereby the three dominant perspectives at play are "product performance," "manufactur-

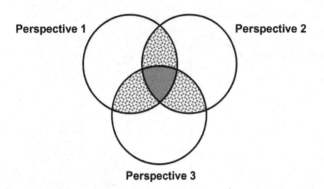

FIGURE 1.10

Identifying engineering solutions by including different perspectives to optimize decisions.

ability," and "product reliability." Engineers with the *product performance* perspective focus on defining a product that properly encloses and protects electromechanical components using a cosmetically appealing package. They approach the design problem with ideas about size, cost, performance, and aesthetics. Engineers with the *manufacturability* perspective are concerned with defining methods for assembly involving the various components and subassemblies that make up the product. They approach the design problem with questions and concerns about consistency of incoming materials and components, fastening methods, and manufacturing costs.

Engineers with the *product reliability* perspective look to see which materials and mechanical designs best lend themselves to the specified user environment. They approach the design problem with concerns about design verification, component variances, and assembly process controls. A healthy interplay between these perspectives early in the design development process can go a long way toward defining an optimum solution that meets the performance, cost, and schedule expectations for the product.

In contrast, consider the impact on this product design process if the product reliability perspective is not consulted or considered early in the design process. Consider the impact on the schedule when learning late in the design and development process that an adhesive joint between two plastic parts is not strong enough to withstand forces typically encountered during customer use, and that the assembly tools have to be redesigned to produce a stronger joint. Consider the impact on manufacturing costs when it is determined during high-volume production that components need to be subjected to 100 percent inspection to catch manufacturing process flaws before the components can be approved for use in production. Consider the impact on product life cycle costs or overall launch success if it is learned that critical mechanical and electrical components are failing prematurely when the product is in the hands of the customer.

Will inclusion of the materials engineering perspective help anticipate such problems, and can cost and schedule risks actually be contained? To the degree that suboptimum materials selection and poor control over the variation of material properties is a culprit in preventing product success, the answer generally is yes.

1.6 TYPES OF PRODUCT DEVELOPMENT PROJECTS

In general, there are four types of development projects. The amount of materials engineering support required during development depends on the specific type of product development project.

Incremental improvements to existing products, which involve adding or modifying features to keep the product line current and competitive. A slight change to a product to eliminate minor flaws is an example of this type of project.

Derivatives of existing product platforms, which extend an existing product platform to better address familiar markets with new products. Examples are automobiles and mobile phones. For automobiles, designers use new body styles with common subassemblies (e.g., chassis, engine, and transmission). The primary functional characteristics are the same between different models, but the look and feel are different. Mobile phones are similar in that common electronics subassemblies are used with different software and phone styling in order to appeal to various types of phone users.

New product platforms, which involve the creation of a new family of products based on a new, common platform, while addressing familiar markets and product categories. Examples are LCD and plasma televisions, hybrid automobiles, and wireless landline telephones. In each case, the new product requires a different set of technologies compared to the cathode ray tube (CRT) television, the internal combustion–only automobile, and the wired telephone, all of which are older families of products.

Fundamentally new products, which involve radically different product or production technologies for new and unfamiliar markets. These projects involve more risk than the other types of projects. Examples of fundamentally new products are the portable music player (Sony Walkman), mobile phones (Motorola), personal digital assistant (Apple Newton), facsimile machine, and television videocassette recorder (Sony Betamax and JVC VHS).

In general, a fundamentally new product will require the most materials engineering support in terms of time and money; an incremental improvement project will require the least. The other two types of projects will fall in between.

1.7 COMPANIES APPLYING THE MATERIALS ENGINEERING PERSPECTIVE

The materials engineering perspective is applicable to all products whether they are assemblies, subassemblies, components, or materials. However, not every aspect of the product development process is applicable to every company. The specific aspects of product design and manufacturing of concern, and the decisions that require the materials engineering perspective, depend on the type of company making the product.

In general, three types of companies manufacture products. The distinction between the categories is based on the level of responsibility that a company has for developing the design requirements for its product and the level of responsibility it has for the design of the product. It is important to make the distinctions, because all the decisions and considerations discussed in this book do not apply to each category of company. The three types of companies are referred to as Type I, Type II, and Type III. The decisions that they face and the application of the materials engineering perspective are described in detail in Chapters 7 through 11. The products sold by Type I, II, and III companies will be referred to as Type I, II, and III products, respectively.

Type I. A Type I company develops the design requirements for its products, which include assemblies, subassemblies, components, and materials used to fabricate components, join components, and form in-process structures. Usually, the same company designs and manufactures the product based on the design requirements. However, there are some exceptions where companies will develop the design requirements and have another company design or manufacture the product. Examples of Type I companies are manufacturers of automobiles, telephones, airplanes, motors, pumps, electronics components (e.g., resistors, capacitors, integrated circuits), home appliances, hand and power tools, gauges, computer memory devices, laboratory glassware, materials (e.g., adhesives, plastics, metals), and kitchen and bathroom faucets.

Type II. A Type II company designs and manufactures a component or subassembly for a single customer based on design requirements developed by the customer. The customer is a Type I or another Type II company that uses the component or subassembly within its own product. Type II companies typically design and manufacture products for more than one customer; and they specify the design requirements for subassemblies and components used within their products.

Type III. A Type III company does not have any design responsibility for the product it manufactures. Instead, a Type III company manufactures its products based on designs developed by client companies, which are Type I and

Type II companies. While the people at Type III companies may provide some input into the design of the product, the client ultimately makes the design decisions. Examples of Type III companies include those that make die cast metal components, stamped or machined metal components, injection molded plastic components, and electronic assemblies and subassemblies.

There are situations when Type III companies rely on other Type III companies to help fabricate a subassembly or component. This situation occurs when a Type III company with a single manufacturing competency uses another Type III company with a different, single manufacturing competency to help fabricate the subassembly or component. An example is a painted metal component; the company responsible for forming the metal component hires a company to paint the component after it has been formed.

1.8 COSTS TO GAIN MATERIALS ENGINEERING KNOWLEDGE

There are costs associated with including the materials engineering perspective and seeking to gain materials engineering knowledge for making better and faster decisions. The costs are related to identifying potential materials and manufacturing processes that can be used, testing the materials to verify their performance and reliability, evaluating the capability of suppliers to control the variation of the materials used in their products, and manufacturing process development. However, the costs and effort required are usually not large compared to the other aspects of product development. That is, the materials engineering efforts typically do not turn into big research projects.

The cost of obtaining materials engineering knowledge is an investment required for developing a successful product and is offset by the benefits associated with making better and faster design decisions. The benefits are as follows:

- Faster product development and faster time to market
- Reduced materials costs
- Meeting performance and reliability requirements
- Higher manufacturing yields
- Reduced manufacturing costs

All the benefits enable a product to satisfy the customer's wants and needs, make a good profit, and establish a good market share.

The relative amount of materials engineering resources required will depend on the type of project. Incremental improvement projects may require little or no additional resources, depending on the changes being made. A derivative product will require materials engineering resources if there are any changes in the design requirements compared to previous versions of the product. For new platform and fundamentally new products, it is critical that materials engineers are included.

The relative amount of materials engineering resources required will also depend on the type of manufacturing company. Type III companies require the least because they have no design responsibility.

1.9 COSTS OF BAD MATERIALS ENGINEERING DECISIONS

Selecting suboptimum materials and exerting poor control over variations of the materials' properties can be costly. Every year, manufacturing companies spend billions of dollars and millions of labor hours addressing problems such as failed product tests, poor supplier quality, poor manufacturing quality, and poor product reliability. The effects of such problems are delayed product launch, field failures, poor customer satisfaction, and poor sales. Each problem can result in unplanned costs or unrealized sales. Furthermore, addressing problems diverts engineering resources from other projects, resulting in delays for other products. All these costs often go unmeasured, but can have a profound impact on a company's profitability and competitiveness.

Selecting materials that result in a product with inferior reliability can be catastrophic to a product's competitiveness and total cost to produce. Conversely, selecting materials that provide reliability well beyond a reasonable safety factor will result in extra, unnecessary costs, which must either be passed on to the customer or absorbed as reduced profits.

When making product design and manufacturing decisions, the total cost of any option must be considered. The costs considered must include those associated with resolving problems related to the materials, not just the cost per unit of the material, component, or subassembly being purchased.

Interestingly, many companies accept design and manufacturing problems as a normal part of business. Furthermore, companies frequently struggle through problems that are related to the materials, and spend more time and money solving them than is necessary to resolve the problems. What these companies do not realize is that many problems can be prevented. Furthermore, when problems with the materials do occur, they can be solved more quickly and with potentially less cost if the appropriate engineers are involved. In some cases, the materials engineering perspective is the only appropriate one.

1.10 THE REMAINDER OF THE BOOK

The rest of this book is divided into three sections. Chapters 2 and 3 discuss the information required to select the materials used in components, joints, and in-process structures. They also review the evaluations necessary to verify that the materials selected will meet the performance, reliability, and cost requirements of the components, joints, and in-process structures.

Table 1.1 Phases of Product Development

	Phase 0: Planning	Phase 1: Concept development	Phase 2: System-level design	Phase 3: Detail design	Phase 4: Testing and refinement	Phase 5: Production ramp-up
Marketing	■ Articulate market opportunity ■ Define market segments	■ Collect customer needs ■ Identify lead users ■ Identify competitive products	■ Develop plan for product options and extended product family ■ Set target sales price point(s)	■ Develop marketing plan	■ Develop promotion and launch materials ■ Facilitate field testing	■ Place early production with key customers
Design	■ Consider product platform and architecture ■ Assess new technologies	■ Investigate feasibility of product concepts ■ Develop industrial design concepts ■ Build and test experimental prototypes	■ Generate alternative product architectures ■ Define major subsystems and interfaces ■ Refine industrial design	■ Define part geometry ■ Choose materials ■ Assign tolerances ■ Complete industrial design control documentation	■ Reliability testing ■ Life testing ■ Performance testing ■ Obtain regulatory approvals ■ Implement design changes	■ Evaluate early production output
Manufacturing	■ Identify production constraints ■ Set supply chain strategy	■ Estimate manufacturing cost ■ Assess production feasibility	■ Identify suppliers for key components ■ Perform make–buy analysis ■ Define final assembly scheme ■ Set target costs	■ Define piece-part production processes ■ Design tooling ■ Define quality-assurance processes ■ Begin procurement of long-lead tooling	■ Facilitate supplier ramp-up ■ Refine fabrication and assembly processes ■ Train workforce ■ Refine quality-assurance processes	■ Begin operation of entire production system
Other Functions	■ Research: Demonstrate available technologies ■ Finance: Provide planning goals ■ General management: Allocate resources	■ Finance: Facilitate economic analysis ■ Legal: Investigate patent issues	■ Finance: Facilitate make–buy analysis ■ Service: Identify service issues		■ Sales: Develop sales plan	

(*Source:* Adapted from Ulrich and Eppinger, 2004; reprinted with the permission of McGraw-Hill.)

In order to select the optimum materials and control their properties, it is necessary to understand the things that affect the properties and how to control them. Chapters 4 through 6 provide this information. First, there is a discussion about the relationship between the properties of a material, its composition and microscopic structure, and the manufacturing processes to which it has been exposed; this discussion covers the science of materials. Chapter 5 reviews manufacturing processes and provides a discussion of the elements of a manufacturing process that must be controlled in order to control the properties of the materials in the item being produced. Finally, Chapter 6 discusses the degradation of materials when they are exposed to various conditions, the effects of the degradation on the properties of materials, and how to evaluate the degradation.

Product development can be divided into the following six phases (Ulrich & Eppinger, 2004):

1. Planning
2. Concept development
3. System-level design
4. Detail design
5. Testing and refinement
6. Production ramp-up

Table 1.1 shows an outline of these different phases and the tasks within each. After product development, there is the production phase.

Chapters 7 through 11 discuss each of the elements of the first five phases of product development that has an impact on the selection or control of the materials used in a product. The discussions consider the information provided in the first two parts of the book. Chapter 11 discusses the materials engineering perspective for elements of the production phase for a product. Finally, Chapter 12 discusses strategies for reducing the risks that are encountered during product development.

REFERENCES

Nevins, J.L., and D.E. Whitney, *Concurrent Design of Products and Processes: A Strategy for the Next Generation in Manufacturing*, McGraw-Hill, 1989.

Ullman, D.G., *The Mechanical Design Process*, Third Edition, McGraw-Hill, 2003.

Ulrich, K.T., and S.D. Eppinger, *Product Design and Development*, Third Edition, McGraw-Hill, 2004

Pahl, G., and W. Beitz, *Engineering Design: A Systematic Approach*, Springer, 1996.

Design Requirements

2.1 INTRODUCTION

As discussed in Chapter 1, products can be broken down into a set of product elements (i.e., components, joints, and in-process structures). Each of these product elements contributes to a function or feature of the product. The performance of each product element influences the performance of the product. Furthermore, the reliability of the product depends on the reliability of the product elements. If a product element fails, then some function or feature will degrade or fail. Finally, the cost to produce a product depends on, among other things, the costs to make or purchase the product elements.

The performance, reliability, and cost of a product element depend on its physical construction and the properties of its constituent materials. The physical construction of a product element refers to its shape and dimensions. It also includes the manner in which multiple materials are incorporated into a product element. For example, some components are composed of a base material with one or more coatings applied over the surface, as shown in Figure 2.1. The physical construction of a product element affects its characteristics and behaviors during use. For example, the physical construction of a product element influences the distribution of mechanical loads along the product element or the manner in which electricity or heat flows through it.

The product elements within a product must be designed so that they enable the product to satisfy all of its requirements, which includes performance, reliability, and cost. The entire set of requirements that must be satisfied comprises the *design requirements*, which are defined during product development. All elements of the design requirements are discussed later in this chapter.

The design requirements for a product are the basis for the design requirements of the product elements. Once the product elements' design requirements have been defined, it is possible to identify, evaluate, and select the materials that can be used for each product element.

There will be design trade-offs between the physical construction and the materials that can be used for a product element. Using a specific physical

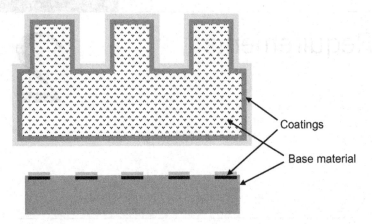

FIGURE 2.1

Schematics of components composed of a material with coatings applied over the surface.

construction can affect the range of materials that can be used for a product element. Alternatively, trying to use a specific material for a product element may constrain its physical construction. Ideally, a product element's physical construction and materials are optimized to provide the required performance and reliability at the lowest cost.

This chapter discusses how the design requirements for product elements are derived from the design requirements of the product. For the sake of the discussions that follow, we assume that a product that is being considered consists of more than one subassembly. Obviously, there are many products that have no subassemblies and that are composed of only product elements or that are a component. Examples are screws, electrical resistors, pencils, bottles, scissors, screwdrivers, and hammers. The discussions that follow can be extended to products like these.

2.2 DEVELOPING DESIGN REQUIREMENTS

The origin of a product's design requirements depends on whether the product is designed by a Type I or Type II company. The design requirements for a Type I product are based on the wants and needs of the intended customer. These wants and needs must be identified by the company making the product, and they are often communicated by the target customer in nontechnical and sometimes vague terms. For example, a product should not be too heavy, should be easy to open, or should look "high-tech." Design teams must convert these wants and needs to technical design requirements. For example, the mass of a product must be 0.5 to 0.7 kg, the force required to open a product should be less than 5 newtons, or a surface must reflect at least 80 percent of incident light.

Once the customer's wants and needs have been identified, the design team converts them to engineering requirements for the product. These engineering requirements become parts of the design requirements for the product. Successfully converting the customer's wants and needs to meaningful engineering requirements necessitates good communication between the marketing and engineering groups within a company. An engineering technique that is useful for helping design teams convert customer wants and needs to engineering requirements is quality function deployment, which is discussed in Chapter 7.

For a Type II product, the company ordering the product provides the Type II company with a set of design requirements. The Type II company designs its product based on these design requirements, which are already in technical terms. As discussed in Chapter 1, the client company, which can be either Type I or Type II, will use the item in its product.

After the design requirements for a product have been defined, the design team develops, evaluates, and selects product *design concepts*, which are descriptions of the product's physical form. After this the design team develops design concepts and defines design requirements for the subassemblies within the product. Subassemblies must be designed so that they satisfy the design requirements of the product. Finally, the design team develops design concepts and defines design requirements for the product elements within the subassemblies. The product elements must be designed so that they satisfy the design requirements of the subassemblies. A flowchart for this whole process is shown in Figure 2.2.

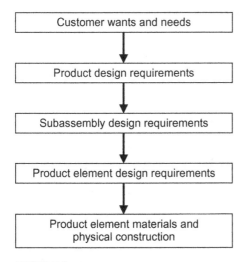

FIGURE 2.2

Flowchart for the design process.

Developing concept designs for a product, its subassemblies, and their product elements is an iterative process. Design teams evaluate the risks associated with the concept designs and their design requirements, and then they make modifications as needed to increase the likelihood of the product's success. This process is discussed in more detail in Chapters 7 through 10.

There can be more than one tier of subassemblies within a product—that is, subassemblies within subassemblies—so that there are sub-subassemblies, sub-sub-subassemblies, and so on. A schematic of this concept is shown in Figure 2.3. Figure 2.3(a) shows the highest-level subassemblies within the product, subassemblies 1, 2, and 3. Figure 2.3(b) shows two subassemblies within subassembly 3.

Each tier of subassemblies must be designed so that it satisfies the design requirements of the subassemblies in which it resides. For the product shown in Figure 2.3, subassemblies 3a and 3b are designed based on the design requirements for subassembly 3.

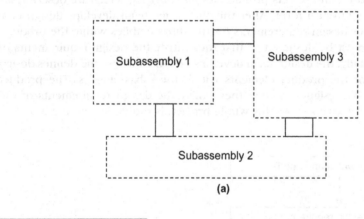

(a)

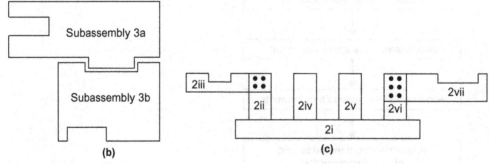

(b)

(c)

FIGURE 2.3

A product and its subassemblies. (a) Complete assembly, (b) subassembly 3 and its subassemblies, and (c) subassembly 2 and its product elements.

For ease of discussion, each subassembly, regardless of its level, will be referred to as a subassembly.

The following sections discuss the design requirements for products, subassemblies, and product elements. Five products will be used as examples. These products are an automobile, a cordless telephone, a motor, an industrial oven, and a cooking skillet. These products, and their subassemblies and product elements that will be used for the discussions, are shown in Figure 2.4.

(a)

(b)

FIGURE 2.4

Example products, subassemblies, and product elements: (a) Automobile.
(b) Cordless telephone. (*Continued*)

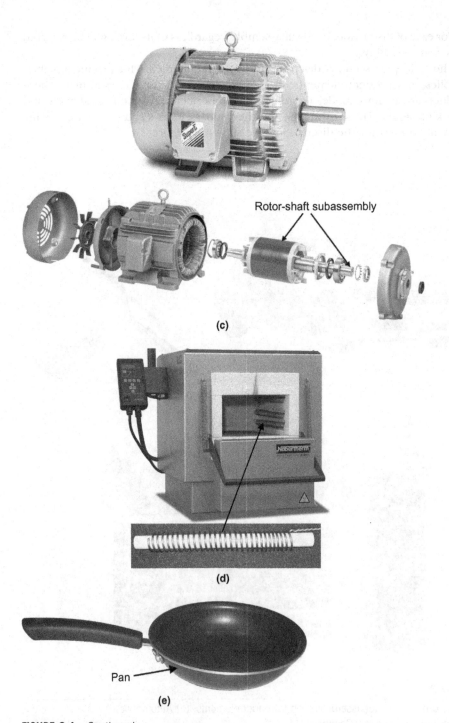

Rotor-shaft subassembly

(c)

(d)

Pan

(e)

FIGURE 2.4 *Continued*

(c) Industrial oven. (d) Motor. (e) Skillet.

2.3 **PRODUCT DESIGN REQUIREMENTS**

Many products have groups of "customers" in addition to those that will be purchasing the product. These other customers are organizations that impose requirements, and although they will not be purchasing the product, it is necessary to satisfy their requirements. Such customers include the following:

- Manufacturing organizations that will be participating in the production of the product
- Industry organizations that set standards covering areas such as product performance, reliability, and safety
- Government organizations that regulate in a variety of areas such as product safety, manner in which products can operate, and the substances that can be used in products
- Legal organizations inside and outside of companies that oversee and control the use of intellectual property

It is important to be aware of all of the relevant nonpaying customers. Not doing so can lead to serious consequences, like a product being banned from sale or being recalled.

The general categories of product design requirements are as follows:

1. Performance requirements
2. Reliability requirements
3. Size, shape, mass, and style requirements
4. Cost requirements
5. Manufacturing requirements
6. Industry standards
7. Government regulations
8. Intellectual property requirements
9. Sustainability requirements

Each of these categories is discussed in more detail later in this section.

This section applies to Type I and Type II products that are assemblies or subassemblies. Type I and II products that are components will be addressed in the section on product element requirements. What follows is a description of each of the nine categories of product design requirements. More details about developing requirements for products are provided in Ulrich and Eppinger (2004).

2.3.1 **Performance Requirements**

The performance requirements describe the functions and features of a product. This involves assigning measurable target values for each performance attribute associated with a particular function or feature. It is important to identify attributes that are measurable; otherwise, it will be very difficult to objectively verify that the requirement is being met.

Some examples of the performance requirements are shown in Table 2.1 for the example products shown earlier in Figure 2.4. The XX's shown in the table take the place of the values that a design team would assign to each particular attribute. For an automobile, the design team would specify the time, measured in seconds, required for the automobile to accelerate from 0 to 100 kilometers per hour.

Table 2.1 Selected Performance Requirements of Example Products	
Product	**Performance Requirements**
Automobile	■ Acceleration: Time from 0 to XX kilometers/hour (seconds) ■ Handling: Minimum turn radius at a specified speed (meters) ■ Fuel economy: Driving distance per unit volume of gasoline (km/liter) ■ Safety features: Air bag deployment time (seconds), braking distance at a specified speed (meters); visibility (glass light transmission, index of refraction) ■ Comfort: Road and engine noise in passenger compartment (decibels)
Cordless telephone	■ Transmission and reception power: XX milliwatts ■ Screen readability: Size of characters (mm), brightness (lumens) ■ Reception range: XX meters ■ Battery: XX ampere-hours/cm^3
Motor	■ Output power: XX watts ■ Starting torque: XX newton-meters ■ Full load torque: XX newton-meters ■ Speed: XX revolutions per minute ■ Noise level: XX decibels at a specific distance ■ Control: On/off, electronic variable control
Industrial oven	■ Exterior size: Length × width × height (cm) ■ Capacity: Length × width × height (cm) ■ Maximum operating temperature: XX°C ■ Temperature uniformity: Temperature variation (°C) ■ Temperature control: Set point or automated ■ Heating efficiency: Energy required to heat up the chamber (watts/cm^3)
Cooking skillet	■ Heating uniformity: Temperature variation (°C) ■ Cooking surface properties: Nonstick or bare metal ■ Maximum use temperature: XX°C ■ Handle insulation: XX watt/m°C

The specific value of the acceleration depends on the wants and needs of the intended customer. Similarly, the target values for the automobile's remaining attributes (which go well beyond the list that is shown in Table 2.1) would be specified based on the wants and needs of the intended customer. The performance requirements for the mobile phone, motor, industrial oven, and skillet contain the same type of information for the attributes that are specific to each product.

The values assigned to each performance attribute correspond accurately to customer wants and needs. If they are not, then the product will end up being over- or underdesigned. In the former case, the product will have unnecessary extra costs because the product was designed to meet requirements that exceeded the customer's wants and needs. In the latter case, the product will not have the desired performance to satisfy the customer's wants and needs. Thus, it is worthwhile to invest the time and resources to properly convert the customer's wants and needs to engineering requirements.

Notice that none of the performance requirements for any of the products shown in Table 2.1 make any mention of the materials used in the products. However, as will be shown, a thorough consideration of all of the performance requirements of a Type I assembly or subassembly is crucial to properly select the materials that will be used to make the product elements.

2.3.2 Reliability Requirements

Reliability refers to a product's ability to perform as specified over a specific use period and under a given set of use conditions. The use period may be measured in different ways such as total ownership time, total on time, or total miles used. The use conditions consist of the mechanical and electromagnetic forces and the thermal, chemical, biological, electrochemical, and radiation environments to which a product is exposed during handling, shipping, and customer use. Examples of different use conditions are as follows:

- *Mechanical:* Static, dynamic, or cyclic loads; impact; rubbing
- *Electromagnetic:* Applied voltage, current, magnetic fields
- *Thermal:* Elevated temperatures; repeated cycling between temperature extremes
- *Chemical:* Gases (e.g., oxygen, nitrogen, or chlorine), liquids (e.g., solvents, acids, or bases)
- *Biological:* Body fluids
- *Electrochemical:* Metals near saltwater or acids
- *Radiation:* Ultraviolet light; radiation in nuclear reactors

The use conditions are associated with the (1) functionality of a product and (2) environment in which the product operates. The mechanical load on a motor as it moves another device and the electricity that passes through an electrical circuit are examples of use conditions associated with product functionality.

Examples of environmental exposure are the use of a motor in the presence of corrosive liquids and the use of an electrical circuit in a high-temperature environment. For products with subassemblies, the environmental conditions may be different from subassembly to subassembly. For example, in an automobile the maximum exposure temperature for a subassembly located in the engine compartment is much hotter than for a subassembly located in the passenger compartment.

Reliability is a concern because the materials that make up a product element can degrade as a result of exposure to the use conditions. The degradation causes changes in the properties of the materials, which can lead to a product element no longer meeting its design requirements, resulting in loss of performance of the product containing the product element.

When defining the use conditions, design teams may have to consider minor abuse and misuse.

Minor abuse refers to exposure to conditions that are somewhat excessive, the determination of which is subjective. Examples of minor abuse include dropping a telephone from 1-meter height onto concrete, driving an automobile over potholes, and small chemical spills onto a motor. Many products are expected to withstand a certain amount of minor abuse by the customer without failing to meet its performance requirements.

Misuse refers to exposure to conditions after which it is unrealistic to expect a product to function properly. For example, it is unrealistic for a customer to expect a cordless telephone to function properly after using it to hammer a nail into a wall or dropping it in a toilet.

A product that ceases to perform as designed is said to have failed. Failure to perform as specified involves the following two situations:

1. A particular function or feature degrades below its minimum performance requirements. The product may still function, but not as well as it did before the degradation occurred. An example is an oven taking too long to heat up or the buttons on a telephone requiring increased force for actuation.

2. A particular functionality is completely lost without warning. Sometimes, these failures are inconvenient, as for an electronic device that will not turn on or an automobile that will not start. In other cases, the loss of functionality is catastrophic, like when an airplane loses rudder control or a ladder collapses. Also, the failure may result in unexpected behavior of the product, resulting in harm to the user, such as when an electrical circuit shorts out and starts a fire.

Unexpected failures that occur during the normal, expected life of a product generally are due to the following five engineering problems:

1. *Inaccurate specification of the use conditions and reliability requirements for a product, its subassemblies, or product elements.* This occurs when a

design team does not thoroughly and accurately identify and quantify the performance and reliability requirements for a product, its subassemblies, or its product elements. Any of these can result in one or more product elements that are underdesigned for the application. The earlier in the design process that poorly defined requirements are made, the worse the problem will be. Underdefined product requirements will certainly result in underdesigned product elements, regardless of the skill of the engineers designing the product elements.

2. *Poor design.* This includes disregarding agreed-on performance and reliability requirements for a product, its subassemblies, or its product elements; not considering industry standards or government regulations; designing a product element with inadequate physical dimensions for withstanding exposure to the use conditions; and selecting suboptimum materials for a product element.

3. *Poor control of the variation of product elements' materials proporties, shape, and size.* Even if the optimum materials are selected, too much variation of their properties and features can lead to product elements that do not have the required properties and therefore do not have the required reliability.

4. *Materials defects.* Defects in a material can compromise its properties to the point that it cannot withstand the exposure to the use conditions without substantial degradation. When this occurs, the affected product element fails to function as designed. Defects typically originate during manufacturing process and are discussed in Chapters 4 and 5.

5. *Product overstress.* Improper use of a product by a customer can cause stresses on a product element that exceed its design limits. Because overstress is a result of inappropriate use of the product, it is beyond the realm of design engineers. We will not be concerned with this item any further.

Ultimately, a product fails unexpectedly when there is an unexpected failure of one or more of its product elements.

Specifying the reliability requirements for a product involves identifying and, when possible, quantifying each use condition to which the product will be exposed. For example, a use condition for a cooking skillet might be exposure up to 250°C. Furthermore, design teams must identify the total exposure to each specific use condition that the product can withstand and still function with the specified performance. For the skillet, the total exposure to 250°C might be 5000 hours, which correlates to over 10 years of use if it is assumed that the skillet is used less than 2 hours a day.

Table 2.2 lists some of the product use conditions for the products discussed in the section on performance requirements. These conditions consider normal use and minor abuse, not misuse.

Table 2.2 Selected Products' Use Conditions

Product	Use Conditions
Automobile	■ Vibration: Up to XX mm displacement, XX m/s velocity, XX m/s^2 acceleration ■ Minimum and maximum temperatures: XX to YY°C ■ Humidity: Up to XX% relative humidity ■ Ultraviolet sunlight: XX mJ/cm^2 ■ Saltwater: Splashing from saltwater on road
Cordless telephone	■ Dropping: XX falls onto concrete from XX m height ■ Button pushing: XX times with YY N force ■ Maximum temperature: XX°C ■ Humidity: Up to XX% relative humidity ■ Contact with hand oils, food, and beverages
Motor	■ Torque: XX newton-meters applied by the motor load ■ Maximum temperature: XX°C ■ Vibration: Up to XX mm displacement, XX m/s velocity, XX m/s^2 acceleration ■ Moisture: Occasional splashing ■ Corrosive gases ■ Explosive gases and dust
Industrial oven	■ Minimum and maximum external temperature: XX to YY°C ■ Humidity: Up to XX% relative humidity ■ Mass of loads placed inside: Up to XX kg ■ Power surges: Up to XX volts ■ Corrosive gases
Cooking skillet	■ Maximum temperature: XX°C ■ Scraping with metal objects ■ Contact with detergents, food, hand oil ■ Dropping

2.3.3 Size, Shape, Mass, and Style Requirements

The size, shape, mass, and style requirements for a product are dictated by its functionality, ease of use, ability to fit into a particular space, and intended aesthetic appeal to customers. Style also includes shape, as well as color and surface texture. Table 2.3 lists some size, shape, mass, and style requirements for the example products.

2.3.4 Cost Requirements

The price at which a company thinks it can sell its product and the desired profit from each sale will set the requirements on the maximum allowed costs to design and manufacture the product. The cost to design a product includes the costs associated with the following:

Product	Size, Shape, Mass, and Style Requirements
Automobile	■ Size: XX cm × XX cm × XX cm (length × width × height) ■ Mass: XX grams ■ Shape: Sedan-like ■ Color: Metallic silver, black, or maroon
Cordless telephone	■ Size: XX cm × XX cm × XX cm (length × width × height) ■ Mass: XX grams ■ Color: Shiny black ■ Surface: Smooth
Motor	■ Size: XX cm × XX cm × XX cm (length × width × height) ■ Mass: XX grams ■ Color: Light blue
Industrial oven	■ Exterior Size: XX cm × XX cm × XX cm (length × width × height) ■ Capacity: XX cm × XX cm × XX cm (length × width × height) ■ Surface: Brushed metal
Cooking skillet	■ Mass: XX grams ■ Diameter of pan: XX centimeters ■ Height of pan: XX centimeters ■ Handle length: XX centimeters ■ Handle shape: Tapered ■ Handle color: Black

Table 2.3 Selected Products' Size, Shape, Mass, and Style Requirements

- Engineering personnel
- Design tools such as computer-aided design and finite element analysis
- Making product prototypes for testing and evaluation
- Product testing
- Materials characterization and materials reliability testing

The cost to manufacture a product includes the costs for the following:

- Purchasing materials, components, and subassemblies used to build the product
- Manufacturing components and subassemblies used within the product
- Assembling the product from the components or subassemblies
- Testing to verify that a product meets the design requirements
- Packaging and shipping the product to customers and stores
- Providing warranty and other product support to customers

The total allowable costs to manufacture a product will constrain the allowable costs for the materials, components, subassemblies, and manufacturing processes used to make the product. None of these take into consideration the costs associated with problems such as failed product tests during product development and poor manufacturing yields.

2.3.5 Industry Standards

Industry standards are documents that contain requirements that have been agreed on by groups of companies and people working in specific industries or on specific types of products. The standards address product performance, safety, reliability, and the methods for evaluating product performance, reliability, and safety.

There are organizations within particular industries and across industries that oversee the development and revision of standards. Some of the organizations that write standards and the products to which the standards apply are shown in Table 2.4. This is a fraction of the standards-writing organizations that exist worldwide. Some of the organizations listed may seem obscure. However, for most of these organizations, the total sales of the products that they oversee are substantial. All of the organizations shown have websites, and most of them have search engines for finding relevant standards. A larger list of standards-writing organizations is given in the *ASM Handbook* (1997). Some of the applicable standards for the example products used in this chapter are shown in Table 2.5.

2.3.6 Government Regulations

Government regulations place requirements on things such as the manner in which a product operates, product safety, and substances that can and cannot be used in the product. Table 2.6 lists some of the agencies in the United States and Europe that write regulations that apply to products. Other countries have their own agencies that write regulations.

Design teams must be knowledgeable of and understand the applications of the government regulations that apply to their products. Failure to comply with existing regulations can result in a product being banned from sale. Table 2.7 lists some of the government regulations that apply to the example products.

2.3.7 Intellectual Property

Intellectual property is a product of an individual's intellect and includes patents, trade secrets, copyrights, and trademarks. A patent is a right given by a government to an inventor (or inventors). The government gives the inventor the sole right to make, use, or sell a product covered by the claims of an issued patent to the inventor(s). The invention must be a process, machine, manufacture, or composition of matter. For a patent to be issued, the invention must meet the government's definition of novel, useful, and nonobvious.

The aesthetic design of a product can also be protected by a design patent, preventing others from using a similar product's nonfunctional styling. If the inventor works for a company, the inventor has usually assigned the rights to the

Table 2.4 Selected Industry Standards-Writing Organizations and Applicable Products or Industries

Organization	Applicable Product or Industry
American Boat and Yacht Council	Safety standards for boat building and repair
American Gear Manufacturers Association	Gears
American National Standards Institute	Covers a wide range of products in different business sectors
ASME International	Boiler and pressure vessels Nuclear components
Association for the Advancement of Medical Instrumentation	Medical devices
ASTM International	Specification and testing of a wide range of materials for a wide range of applications.
Institute of Electrical and Electronics Engineers	Electronics and electrical devices
International Electrotechnical Commission	Electrical, electronic, and related technologies
International Organization for Standardization	Covers a wide range of products and applications
IPC	Electronic circuit board manufacturing and assembly
Metal Powder Industries Federation	Powder metal components
National Electrical Manufacturers Association	Electrical equipment
NSF International	Food, water, and consumer products
SAE International	Aerospace and ground vehicles (e.g., automotive, agricultural, off-road machinery, construction), trucks and buses
The JEDEC Solid State Technology Association	Semiconductor components
Underwriters Laboratories	Fire and explosion safety of electrical products

patent to his or her employer as a condition of employment. The different types of intellectual property are as follows.

A *patent* right is a negative right in that the patent holder can prevent others from practicing any of the claims in the invention; however, it does not give the patent holder the right to practice the invention described in the claims because doing so could infringe on someone else's patent rights or be against

Table 2.5 Selected Industry Standards Applicable to Example Products

Assembly/ Subassembly	Applicable Industry Standards
Automobile	■ SAE J299 Stopping Distance Test Procedure ■ SAE J144 Subjective Rating Scale for Vehicle Handling ■ SAE J850 Fixed Rigid Barrier Collision Tests
Cordless telephone	■ TIA-1083 Telephone Terminal Equipment Handset Magnetic Measurement Procedures and Performance Requirements ■ TIA/EIA-470-C.310 Cordless Telephone Range Measurement Procedures
Motor	■ IEC 60034 Rotating Electrical Machines ■ IEEE 114 Standard Test Procedure for Single-Phase Induction Motors ■ NEMA MG1 Motors and Generators
Industrial oven	■ National Fire Protection Association (NFPA) 86: Standard for Ovens and Furnaces ■ NEMA 250 Enclosures for Electrical Equipment
Cooking skillet	■ NSF/ANSI 51 Food Equipment Materials

Table 2.6 Selected Government Regulatory Agencies

Government Agency	Applications
United States Federal Communications Commission	Products that emit electromagnetic radiation
United States Federal Aviation Administration	Aviation products
United States Nuclear Regulatory Commission	Products that use radioactive materials
United States Occupational Safety & Health Administration	Product safety
United States Consumer Products Safety Commission	Consumer safety
United States Food and Drug Administration	Medical devices
European Union	Many

government policy. An inventor can license or sell the right to use the patent to others.

A *trade secret* is information that (1) derives independent economic value from not being generally known or readily ascertainable to other persons who can obtain economic value from its disclosure or use and (2) is the subject of efforts that are reasonable under the circumstances to maintain its secrecy. Examples

Table **2.7** Selected Products and Applicable Government Regulations	
Assembly/Subassembly	**Regulation**
Automobile	■ Directive 2000/53/EC (European Parliament) End-of life vehicles: Regulates the substances restricted for use in automobiles ■ U.S. Code of Federal Regulation Title 49 Part 571 Federal Motor Vehicle Safety Standards: Regulates motor vehicle safety requirements ■ U.S. Clean Air Act: Regulates the types and amounts of emissions from automobile
Cordless telephone	■ U.S. Code of Federal Regulation Title 47 Telecommunication Part 15 Radio Frequency Devices Council: Regulates the allowable electromagnetic radiation emitted from electronics ■ Directive 2002/95/EC (European Parliament) Restriction of the use of certain hazardous substances in electrical and electronic equipment: Regulates the substances restricted for use in electronic devices
DC motor	■ U.S. Code of Federal Regulation Title 10 Energy Part 431 Energy Efficiency Program for Certain Commercial and Industrial Equipment: Regulates energy conservation requirements for electric motors ■ U.S. Code of Federal Regulation Title 47 Telecommunication Part 15 Radio Frequency Devices Council: Regulates the allowable electromagnetic radiation emitted from electronics
Industrial oven	■ U.S. Code of Federal Regulation Title 47 Telecommunication Part 15 Radio Frequency Devices Council: Regulates the allowable electromagnetic radiation emitted from electronics
Cooking skillet	■ U.S. Code of Federal Regulation Title 21 (Federal Food, Drug and Cosmetic Act regulations): Regulates the use of nonstick coatings ■ Regulation (EC) No. 1935/2004 of the European Parliament and of the Council of 27 October 2004 on materials and articles intended to come into contact with food: Regulates the use of cooking utensil materials

of trade secrets can be formulas for products, methods used in manufacturing, and compilations of information.

A *copyright* protects the embodiment of a work (e.g., a work of art, software, or a performance), not the inventive ideas behind the work.

A *trademark* or *service mark* is a word, phrase, symbol, design, or combination of these that identifies and distinguishes the source of the goods or services of one party from those of others.

One effect of patents owned by other companies is to restrict the implementation of certain functions within a product or restrict the means for achieving the functions and features. By the way, it is possible for a patent for an unrelated type of product to apply to the functions or features that are intended within the product being designed. For the case of a patent covering the implementation of certain functions, the design team must decide to either purchase the right to use the invention covered by the patent or exclude the functionality from the product. For the case of a patent covering the means for achieving certain functions or features, the design team must come up with a new invention for achieving the functions or features, purchase the rights to use the invention covered by the patent, or exclude the functions and features from the product.

It is best to gain the knowledge about patent coverage at the beginning of a design effort. Design teams can conduct patent searches before starting the design of a new product. Resources for researching patents are available on the Internet. Some of the various worldwide patent offices and their websites for researching patents are the following:

- U.S. Patent and Trade Office, *www.uspto.gov*
- European Patent Office, *www.european-patent-office.org*
- Japan Patent Office, *www.jpo.go.jp*

2.3.8 Manufacturing Requirements

Companies that have internal manufacturing capabilities may require that specific manufacturing processes and materials be used to produce the product. For example, these constraints can restrict the product's shape and dimensions so that the existing manufacturing equipment can handle the product as it is being built. Perhaps the manufacturing capabilities may restrict the types of joints that can be used to join components and subassemblies, which will have an impact on the materials that can be used to form the joints.

For a product that is similar to past products, manufacturing constraints such as these are acceptable. In fact, they may be desirable because they provide a certain level of confidence in the design of products that can be produced using familiar processes and materials.

In cases where a new product differs significantly from past products, the constraints of using specific materials and processes may provide a competitive advantage or be a burden. It can be a competitive advantage if the design team is creative enough to develop a product around the manufacturing requirements. However, the requirements are a burden if the existing processes are inherently incompatible with the new product or if the design team does not have the required creativity.

2.3.9 Sustainability Requirements

Sustainability means "meeting the needs of the present without compromising the ability of future generations to meet their own needs." It requires that human activity only utilizes nature's resources at a rate at which they can be replenished naturally. The needed aim of sustainable design is to manufacture products in a way that reduces use of nonrenewable resources, minimizes environmental impact, and relates people with the natural environments.

Sustainable design (also referred to as "green design," "eco-design," or "design for environment") is the art of designing physical objects to comply with the principles of ecological sustainability. Sustainable technologies are those that use less energy, use fewer limited resources, do not deplete natural resources, do not directly or indirectly pollute the environment, and can be reused or recycled at the end of their useful life.

Sustainable design is a general reaction to the rapid growth of economic activity and human population, depletion of natural resources, damage to ecosystems, and loss of biodiversity. It is considered a means of reducing the use of nonrenewable resources and our impact on the environment, while maintaining quality of life, by using clever designs to substitute less harmful products and processes for conventional ones.

The following are some common principles of sustainable design:

- Use low-impact materials. Choose nontoxic, sustainably produced, or recycled materials that require little energy to process.
- Use manufacturing processes that require less energy.
- Make longer-lasting and better-functioning products that will have to be replaced less frequently, reducing the impact of producing replacements.
- Design products for reuse and recycling.
- Use materials that come from nearby (local or bioregional), sustainably managed renewable sources that can be composted when their usefulness has been exhausted.

For example, automobiles and appliances can be designed for repair and disassembly (for recycling) and constructed from recyclable materials such as steel, aluminum, and glass, and from renewable materials (e.g., wood and plastics from natural feedstock).

From the standpoint of product design requirements, sustainability requirements are currently based on whether or not it is important to customers that their purchase of the product is environmentally responsible. Thus, it is actually an aspect of the customers' wants and needs that influences the materials and processes that can be used. This in turn will affect the design of the product and its subassemblies.

2.4 SUBASSEMBLY DESIGN REQUIREMENTS

The process for identifying the design requirements for a product's subassemblies is similar to that described in the previous section for the product. The same design requirements discussed there must be defined for each subassembly in a product. The performance and reliability requirements for one subassembly per product for the example products used earlier are shown in Tables 2.8 and 2.9.

Table 2.8 Selected Subassemblies and Their Performance Requirements

Subassembly	Performance Requirements
Automobile windshield wiper blade	■ Push water off windshield: XX% visibility ■ Support loads from wiper arm: XX kg force
Cordless telephone circuit board	■ Process transmission and reception signals: XX instructions per second ■ Information loss: < XX processing errors ■ Battery power consumption: XX ampere-hours ■ Command processing: redial, volume control, and so on ■ Memory: Store XX phone numbers
Motor rotor-shaft	■ Carry current: XX amperes ■ Transmit torque to load: XX newton-meters
Industrial oven heating element	■ Heat oven up: > XX°C ■ Produce uniform heating: < XX°C variation
Cooking skillet	■ No subassemblies

Table 2.9 Selected Subassemblies and Their Use Conditions

Subassembly	Use Conditions
Automobile windshield wiper blade	■ Maximum temperature: +YY°C ■ Up to XX% relative humidity ■ Ultraviolet sunlight ■ Splashing from saltwater on road ■ Windshield wiper fluid
Cordless telephone circuit board	■ Impact: Drop 2 meters onto concrete ■ Maximum temperature: +YY°C ■ 85% relative humidity ■ Contact with food, beverages, hand oils, lotions, and other household chemicals
Motor rotor-shaft	■ Maximum temperature: +YY°C ■ Cyclic loading: Torsion loads between −XX and +XX N-m ■ Vibration: Up to XX mm displacement, XX ns velocity, XX m/s acceleration ■ Moisture, corrosive liquids, corrosive gases ■ Explosive gases and dust ■ Dust ■ Ignitable fibers
Industrial oven heating element	■ 1500°C ■ Thermal cycling: Between 25°C and 1500°C ■ Corrosive vapors
Cooking skillet pan	■ No subassembly

2.5 PRODUCT ELEMENT DESIGN REQUIREMENTS

The general categories for product element design requirements are the same as those for assemblies and subassemblies. This section discusses each of these categories as they apply to product elements, using the following product elements as examples:

- Automobile windshield wiper blade insert
- Cordless telephone circuit board solder joint
- Motor shaft
- Industrial oven heating element
- Cooking skillet pan

2.5.1 Performance Requirements

Each product element within an assembly or subassembly has a specific set of functions that enable the functionality and performance of an assembly or subassembly. The performance requirements describe the attributes that the product element must have to function as required. The attributes can be described in terms of mechanical, electromagnetic, thermal, optical, physical, chemical, electrochemical, and cosmetic properties.

The performance requirements for the product elements within the example products and subassemblies discussed earlier are shown in Table 2.10. The product elements were shown earlier in Figure 2.4. Each performance require-

Table 2.10 Selected Product Element Performance Requirements

Product Element	Performance Requirements
Automobile wiper blade insert	- Conform to the surface of the windshield: Elastic modulus < XX MPa - Not deform too much: Elastic modulus > YY MPa - Slide easily across the surface of the windshield with low noise: coefficient of friction < X, unitless
Cordless telephone circuit board solder joint	- Conduct electricity between circuit board and electrical component: < XX ohms - Hold component in place on the circuit board: Strength > XX MPa
Motor shaft	- Transfer power from rotor to load: Strength > XX MPa and elastic modulus (rigidity) > XX MPa
Industrial furnace heating element wire	- Heat oven to maximum temperature: Electrical resistance XX–YY ohms)
Cooking skillet pan	- Hold food: Strength > XX MPa and elastic modulus (rigidity) > XX MPa - Uniform heating: Thermal conductivity > XX watts/m · K - Nonstick surface: Surface energy < XX ergs

ment consists of an attribute and the required value for the attribute. The lower and upper bounds of the required property are indicated by X and Y. The value is stated as a minimum, maximum, or range. For example, the wiper blade insert must be flexible enough so that it conforms to the surface of a windshield, but not so flexible that it deforms too much when pressed against the windshield. Also, the insert must slide easily across the surface of a windshield. The flexibility requirements can be described by the elastic modulus of the insert material, which can have a range of values that correspond to acceptable flexibility. The sliding requirements can be described by the coefficient of friction of the insert material, which must be below a maximum value. The same approach to specification of the product element performance requirements is applied to the telephone solder joint, motor shaft, industrial furnace heating element wire, and cooking skillet pan.

There is one significant difference between the design requirements for products and subassemblies and the design requirements for product elements. The performance requirements for products and subassemblies are described only in terms of the functionality that they provide, whereas the performance requirements for product elements are described in terms of material properties (e.g., strength, electrical conductivity, melting-point temperature). This is significant because these material properties, along with the other product element design requirements, will be used as part of the criteria for identifying and selecting materials that can be used to form a product element.

2.5.2 Reliability

The reliability of a product element refers to its ability to function as required over a specific use period when exposed to a specific set of use conditions. The reliability of a product element depends on its materials, the use conditions to which it is exposed, and the response of the materials to the use conditions. The use conditions can cause degradation of the materials to the point that the product element no longer performs as required. The degradation can occur quickly or over a long period of time. In either case, the product element fails once the material degrades to the point at which the product element no longer performs as required. The manner by which materials degrade and the effects of specific use conditions on materials is discussed in Chapter 6.

Within an assembly or subassembly there will be different use conditions from product element to product element. The specific use conditions for a product element are related to the following three items:

1. The *functionality of a product element*. While performing its function, a product element is exposed to mechanical, electromagnetic, thermal, chemical, biological, electrochemical, or radiation conditions. The mechanical forces acting on the tip of a wiper blade are due to the force of the wiper subassembly pushing the blade against the windshield and the motion of the blade across the windshield. The on and off cycling of an industrial oven will expose the

heater element wire to thermal stress fatigue. Also, the high temperatures will expose the heater element wire to oxidizing conditions.

2. The *physical location of a product element within an assembly or subassembly.* The product elements within a product can be exposed to different environmental use conditions depending on their physical location within the product. A product element located within the passenger compartment of an automobile is not exposed to the same temperature and corrosion conditions as a product element located in the engine compartment. The pan of a skillet is exposed to much higher temperatures compared to the handle. The differences in the specific environment require a design team to carefully consider the use conditions for every product element in a product.

3. *Interactions between product elements.* There can be interactions between product elements that are not directly related to how they function. If the materials are incompatible, then the interactions can lead to degradation of one or both of the product element's materials. For example, galvanic corrosion can result when two metallic product elements are in contact in the presence of moisture. Another example is a product element made of a polymer or rubber material that, when exposed to elevated temperatures, releases a gaseous chemical that reacts with other product elements located nearby. When selecting materials, design teams must understand the responses of the different materials to the environmental conditions and the effects of the responses on the other materials in the vicinity.

Example product elements' use conditions are shown in Table 2.11.

2.5.3 Size, Shape, and Mass Requirements

The size, shape, and mass requirements for a product element will have a huge influence on the materials that can be used. Consider a component that must carry five amperes of current without heating up by more than 15°C above the ambient temperature. The electrical conductivity for a component with a 1-mm diameter must be about four times greater than the electrical conductivity for a component that can be 2 mm in diameter. A bicycle frame that has to be 8 kg must have frame tubes made of a lower-density material compared to a 12-kg frame. For a component that needs to support 100 kg, the yield stress for the material in a component that must be 5 mm in diameter has to be much greater compared to the material in a component that can be 10 mm in diameter.

The product element size, shape, and mass requirements for the example product elements are shown in Table 2.12.

2.5.4 Cost

The cost to form a product element or purchase a component depends on the factors that follow.

Table 2.11 Selected Product Elements and Their Use Conditions

Product Element	Use Conditions
Automobile wiper blade insert	<ul><li>Pressure against windshield: XX newtons</li><li>Cyclic motion: Back and forth</li><li>Temperature extremes: XX–YY°C</li><li>Relative humidity: XX%</li><li>Ultraviolet radiation, wiper fluid, saltwater</li></ul>
Cordless telephone circuit board solder joint	<ul><li>Temperature extremes: XX–YY°C</li><li>Relative humidity: XX%</li></ul>
Motor shaft	<ul><li>Torsion loads: XX newton-meters</li><li>Bending loads: XX newtons</li><li>Temperature extremes: XX–YY°C</li><li>Corrosive liquids and gases</li></ul>
Industrial furnace heating element wire	<ul><li>Temperature extremes: XX–YY°C</li><li>Thermal cycling</li><li>Chemical vapors</li></ul>
Cooking skillet pan	<ul><li>Temperature extremes: XX to YY°C</li><li>Scraping surface</li><li>Dropping 1 meter onto tile floor</li><li>Food, detergents, hot water</li></ul>

Table 2.12 Selected Product Element Size, Shape, and Mass Requirements

Product Element	Size, Shape, and Mass Requirements
Automobile wiper blade insert	<ul><li>Length: XX mm</li><li>Shape: Cross-section shape (e.g., T-shaped)</li></ul>
Cordless telephone circuit board solder joint	<ul><li>Shape: Fillet</li></ul>
Motor shaft	<ul><li>Length: XX centimeters</li><li>Diameter: XX centimeters</li></ul>
Industrial furnace heating element wire	<ul><li>Length: XX centimeters</li><li>Diameter: XX centimeters</li></ul>
Cooking skillet pan	<ul><li>Diameter: XX centimeters</li><li>Mass: XX grams</li></ul>

The materials that constitute a product element. As the design requirements for a product element become more restrictive, the range of materials options available to the design team decreases and the cost of the materials that can satisfy the requirements increases. Looking at just performance and reliability, materials that result in low product element performance and reliability will be the least expensive compared to materials that enable greater product element performance and reliability.

The manufacturing processes used to form a product element. Some processes are more expensive than others because of the process complexity, equipment required, skill required to operate the processes, or the number of people required to perform the work. There is not a direct relationship between the complexity and cost of a manufacturing process and the performance and reliability requirements of a component. There are many low-complexity processes used to fabricate high-performance and high-reliability components.

Whether a component is custom made or purchased from an "off-the-shelf supplier." Off-the-shelf components typically cost less than custom-designed components because they are made in higher quantities compared to custom ones. The higher volumes enable off-the-shelf components to be manufacturered at lower prices compared to custom-made components.

The quantity of components or materials being purchased. Typically, as the number of components being purchased increases, the cost per piece decreases. This is also true for materials used to form components, joints, and in-process structures.

Quality problems associated with a component or material. Many companies seek out the lowest cost supplier of components and materials, and sometimes neglect consideration of the potential costs associated with manufacturing quality and product quality problems. The technical capabilities and skill of the supplier of a component or material will have an impact on quality. If the performance and reliability of a supplier's product are poor or inconsistent, then there will be additional, unplanned costs associated with poor manufacturing quality, customer returns, warranty repairs, and product recalls.

2.5.5 Industry Standards

Just as there are industry standards for assemblies and subassemblies, there are also industry standards that address the performance and reliability of product elements. In some cases, a specific standard will discuss assembly, subassembly, and product element requirements. Standards specific to product elements address issues such as the following:

- The size and shape of product elements used for specific applications
- The materials that can and cannot be used for product elements used for specific applications
- The tests required to verify the properties of the materials used to make a product element

The standards-writing organizations discussed earlier also apply to standards for product elements. Table 2.13 lists selected product elements and examples of applicable standards.

Table 2.13 Product Elements and Selected Applicable Industry Standards

Product Element	Applicable Industry Standards
Automobile wiper blade insert	■ None found
Cordless telephone circuit board solder joint	■ IPC 9701A Performance Test Methods and Qualification Requirements for Surface Mount Solder Attachments ■ IPC J-001D Requirements for Soldered Electrical and Electronic Assemblies ■ IPC-A-610D Acceptability of Electronic Assemblies
Motor shaft	■ None found
Industrial furnace heating element wire	■ ASTM F289 Standard Specification for Molybdenum Wire and Rod for Electronic Applications ■ ASTM B76 Standard Test Method for Accelerated Life of Nickel-Chromium and Nickel-Chromium-Iron Alloys for Electrical Heating
Cooking skillet pan	■ NSF/ANSI 51 Food Equipment Materials

2.5.6 Government Regulations

Government regulations regarding components used in assemblies and subassemblies are typically related to requirements on the materials from which components can and cannot be made. The requirements address concerns about the toxicity of the substances used in a product element, the reliability of the materials in specific applications, and the impact of the substances used on the environment after the product has been thrown away. Every country has its own set of regulations.

One example is European Union legislation (Directives 2000/53/EC, 2002/95/EC, 2002/96/EC) that restricts the use of certain substances from certain types of electrical and electronics products. Some of the restricted substances are lead, mercury, and cadmium. For the products covered by the legislation, the amounts of these substances must be below certain levels in the materials that make up individual product elements. Thus, these regulations affect the materials' options available to make individual product elements.

2.5.7 Intellectual Property

There are many patents regarding the design and manufacture of product elements. Just as for an assembly or subassembly, if a patent is found that is applicable to the product element being selected or designed, then the design team has to decide whether to license the patent or engineer the product element to avoid conflict with the patent.

2.5.8 Manufacturing Requirements

Type I and Type II companies may require that specific processes be used for fabricating components and building assemblies or subassemblies. One reason for the restrictions is that a company has internal manufacturing capabilities that must be used, and there is no desire to invest in additional manufacturing capabilities. Another reason is that a company has experience with and is comfortable with product elements fabricated using a familiar manufacturing process, and wants to reduce the risk for problems when developing a new product.

Restrictions on the processes that can be used will restrict the materials that can be used to make product elements because the materials must be compatible with the processes. For example, components that must be joined using a specific welding, brazing, or soldering process must be made out of materials that enable good joints to be formed using the joining process. This may exclude the use of off-the-shelf components from one or more suppliers because the components are made of materials that are incompatible with the process. For a custom component, the restriction may require the use of certain materials in order to form a good joint.

Restricting the manufacturing process to only familiar ones will restrict the options of materials that can be used to form a product element because in many cases a specific set of materials can be used with a particular process. For example, only certain aluminum, zinc, or magnesium alloys can be used with die casting. In some respects, these constraints are acceptable, and may in fact be desirable, because they provide confidence in the design of components, subassemblies, and products produced using processes and materials with which there is experience. In cases when a new product is significantly different than older products, the constraints of using specific manufacturing processes may seem to be a burden.

2.5.9 Sustainability Requirements

The sustainability requirements for the product become the sustainability requirements for the product elements. These requirements restrict the materials that can be used in product elements to materials such as those that can be reused or recycled. The requirements will also restrict the manufacturing processes that can be used to form product elements to processes that do not harm the environment and do not use chemicals and materials manufactured using environmentally unfriendly processes.

REFERENCES

ASM Handbook, Volume 20: Materials Selection and Design, ASM International, 1997.

Directive 2000/53/EC of the European Parliament and of the Council of 18 September 2000 on End-of Life Vehicles.

Directive 2002/95/EC of the European Parliament and of the Council of 27 January 2003 on the Restriction of the Use of Certain Hazardous Substances in Electrical and Electronic Equipment.

Directive 2002/96/EC of the European Parliament and of the Council of 27 January 2003 on Waste Electrical and Electronic Equipment (WEEE).

Ulrich, K.T., and S.D. Eppinger, *Product Design and Development*, Third Edition, McGraw-Hill, 2004.

Selecting Materials

3.1 INTRODUCTION

As discussed in Chapter 1, the materials engineering perspective is concerned with the following issues:

- The performance, reliability, and cost of a product are strongly dependent on the properties of the materials that make up the product
- Selection of materials with properties that enable a product's design requirements to be satisfied
- Controlling the variation of the materials' properties

This chapter provides a brief discussion of the process and considerations for selecting materials. Chapters 5 and 6 provide more detail about why the considerations are important. The remaining chapters explain the application of the materials engineering perspective to various elements of product design and manufacturing.

3.2 MATERIALS SELECTION PROCESS

A flowchart for the materials selection process is shown in Figure 3.1. The process consists of the following steps:

1. Identify product design requirements
2. Identify product element design requirements
3. Identify potential materials
4. Evaluate materials
5. Determine whether any of the materials meet the selection criteria
6. Select materials

The first two items were discussed in Chapter 2. The last four items will be discussed in this chapter and in following ones.

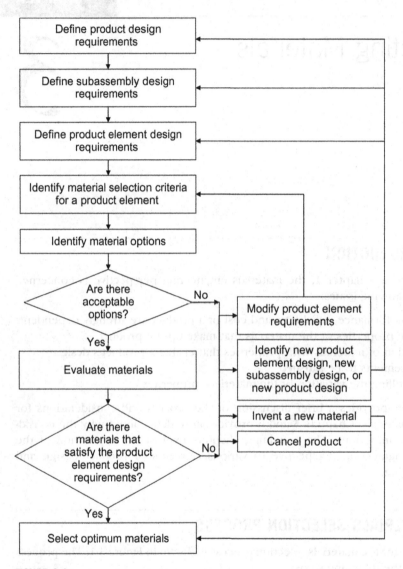

FIGURE 3.1

Materials selection process flowchart.

During the process of identifying and evaluating materials, a design team may determine that there are no materials that can be considered for use for a product element. In this situation, the design team has the following options:

1. Modify the design of the product element.
2. Modify the design of the product or subassembly that directly uses the product element.

3. Modify the design requirements of the product.
4. Invent a new material.
5. Cancel the product.

It is critical that design teams determine whether there are no options as soon as possible in the design process because it will give them the option of modifying the design or design requirements of the product element, subassembly, or product when it is still easy and inexpensive to make changes. Waiting too long will force design teams to consider either trying to invent a new material or canceling the product. Inventing a new material adds cost and risk to the development effort. However, the added cost and risk may be worthwhile if there is an invention that provides the product with a competitive advantage compared to products from other companies. Finally, canceling a product may be undesirable; however, it is preferable to spending time and money developing a product that does not meet the customer's wants and needs.

There is one more option that the flowchart does not show—moving forward with a suboptimum material. This means that the product element will not have the necessary performance or reliability, which reduces the likelihood of having a successful product.

Waiting until all the relevant design requirements have been identified is important because doing so will prevent a design team from pursuing suboptimum materials based on incomplete information. Taking the time to make sure that all the relevant requirements have been identified will increase the chances of selecting optimum materials and enabling a successful product.

3.3 IDENTIFYING POTENTIAL MATERIALS

Once the required material properties have been identified for a product element, it is possible to identify potential materials that can be used for the product element. The other product element design requirements put further constraints on the materials that can be used, often reducing the set of potential materials that can be considered. For example, the European Union Directive 2002/95/EC, Restriction of the Use of Certain Hazardous Substances in Electrical and Electronic Equipment, prohibits the use of lead (Pb) in electronics. Therefore, a company that makes mobile phones for sale in Europe must use lead-free solder paste to form solder joints. The number of potential materials that can be considered for use within a product element decreases as the number of requirements is added and as the requirements become more demanding and complex.

Materials have many different properties, including physical, mechanical, thermal, electrical, chemical, electrochemical, and optical properties. Table 3.1 lists some material properties and defines each one. There are many other properties that have not been listed here. Not every property listed in the table is applicable to every material. For example, a solid metal does not have a viscosity or a

Table 3.1 Basic Materials Properties and Their Definitions

Physical	▪ Density: Mass per unit volume. ▪ Permeability: Amount of gas or liquid that can pass through a material. ▪ Molecular weight: Mass per mole of a substance. ▪ Molecular weight distribution: Statistical distribution of molecular weights in a polymer.
Mechanical	▪ Yield strength: Stress at the onset of plastic deformation during tensile loading. ▪ Ultimate strength: Maximum stress that can be supported before fracture during tensile loading. ▪ Shear strength: Maximum stress that can be supported before fracture during shear loading. ▪ Elastic moduli (Young's, shear, bulk): Ratio of stress to strain during elastic (nonpermanent) deformation. ▪ Hardness: Amount of deformation induced in the surface as the result of an applied indentation load. ▪ Fatigue strength: Relationship between applied cyclic stresses and the number of cycles to failure. ▪ Toughness: Total energy required to cause fracture. ▪ Fracture toughness: Maximum stress that can be applied before catastrophic crack propagation. ▪ Creep resistance: Resistance to deformation while stressed at elevated temperatures. ▪ Coefficient of friction: Ratio of friction force to load for a material moving across a surface. ▪ Viscosity: Shear property of a fluid or paste. Relates to the flow of a fluid or paste.
Thermal	▪ Melting point (T_m): Temperature at which a material changes from solid to liquid. ▪ Glass transition temperature (T_g): Temperature of the transition from solid to very viscous liquid. ▪ Coefficient of thermal expansion: Thermal strain per degree of temperature change. ▪ Thermal coefficient of resistance: Change in electrical resistance per degree of temperature change. ▪ Thermal conductivity: Rate at which heat flows through a material. ▪ Specific heat: Amount of heat required to raise a substance's temperature 1°C.
Electrical	▪ Conductivity: Ability of electrons to flow through a material. ▪ Dielectric strength: Maximum voltage gradient that a material can withstand before breakdown. ▪ Dielectric constant: Ability of a material to resist the formation of an electric field within it. ▪ Magnetic permeability: Change in magnetic induction in response to a magnetizing force.
Chemical	▪ Reaction rate: Rate of chemical reaction at a particular temperature with a specific liquid or gas.
Electrochemical	▪ Electrochemical potential: Voltage that arises between a material and a reference material that are electrically connected in the presence of an electrolyte.
Optical	▪ Refractive index: Ratio of the velocity of light in a vacuum to its velocity in a material. ▪ Transmission: Amount of light of a specific wavelength that can pass through a material.

glass transition temperature. Also, for a particular material, not all its properties are relevant for any particular use of the material.

Identifying potential materials will also include identifying the manufacturing processes used to form a product element. The processes have an effect on the properties of the materials. For example, a component machined out of wrought aluminum will have different properties compared to the same component die cast out of aluminum. Also, there are many manufacturing processes that can be used only with a specific set of materials.

The product element design requirements help design teams focus on considering only those materials that will enable them to meet the design requirements for a product element. In some cases, the design requirements are so restrictive that few or no materials can be identified for use. Too few materials options adds risk to the product development effort if there are concerns that testing and evaluation might reveal that none of the options are viable. The risk is reduced if at least one option is known to have a high probability of meeting all the design requirements.

The cases when a product element's design requirements result in few material options forces design teams to be creative about their designs. For example, a component may have certain strength, electrical conductivity, and cost requirements that can only be satisfied by using a nontraditional combination of materials and certain fabrication methods.

Creative solutions to material selection decisions can provide a competitive advantage if the materials enable a product with better performance, better reliability, or lower cost compared to competitors' products. Sometimes the design requirements for a product element are mutually exclusive, requiring the invention of a new material or combination of materials, which can also provide a competitive advantage.

Options for the materials that can be used to make a product element can be identified based on available published information about them. Generally there is a great deal of material properties information in handbooks, textbooks, journals, standards, and suppliers' technical data sheets; all of these are discussed in Chapter 4.

3.4 EVALUATING THE MATERIALS

Quite often the available published information about a material is incomplete. There may be enough information to consider whether a material is a potential candidate for use, but not enough to select it for the final design. Thus, it is necessary to conduct evaluations to determine whether a material does satisfy all the design requirements of the product element under consideration. When there is more than one option of materials, evaluations provide information that will help identify the material with properties that best meet the design requirements, at the lowest cost.

The evaluation process consists of obtaining the following information about aspects of the materials:

- Properties
- Features
- Variation in properties
- Degradation characteristics
- Ease of use in manufacturing processes
- Total cost to use

Each of these is discussed next and in more detail in Chapters 4 through 6.

Design teams should consider comparable materials from two different suppliers as being two different materials. Even if the materials have the same nominal composition, there will probably be differences in their properties because of (1) the methods and care used to prepare the materials and (2) differences in additives used to modify the material properties.

3.4.1 Material Properties

The available published information about the properties of material is not always complete. Perhaps not all the significant material properties for a particular application are available because they have never been measured or reported. Another reason for insufficient information is that the data available may be single valued, or identified as being "typical." However, there is actually a range of values for any property for a material. It is important to determine the maximum and minimum values for a specific property to determine whether the material will meet a specific design requirement for all samples of the material. Some characterization of the properties of the materials may be required to determine whether a material does indeed meet all design criteria.

3.4.2 Material Features

Material features refers to a material's composition and the microscopic structures within it. As we will discuss in the next chapter, the properties of a material depend directly on its features. Furthermore, some features depend on the manufacturing processes used to form a product element or produce a material used to form a product element. It is important to evaluate material features because it is possible to form a product element that appears to have the correct properties, but its material features are suboptimum, which can lead to reliability problems.

3.4.3 Variation in Material Properties

There will be some variation in the properties of the materials that constitute a product element. Achieving consistent product performance and reliability requires that the variation be within certain limits. As discussed in Chapter 5, variation in the properties of the materials in a product element depends on the materials and manufacturing processes used to make it.

3.4.4 **Degradation Characteristics**

The materials that make up a product element will degrade when they are exposed to the mechanical, electromagnetic, thermal, chemical, biological, electrochemical, and radiation conditions during use of the product. This degradation results in changes in the properties of the material, and sometimes in the failure of the product element and the product. We will discuss this aspect in more detail in Chapter 6.

There is a great deal of materials degradation information in handbooks, textbooks, journals, and sometimes in suppliers' technical data sheets. However, this information is often incomplete, is based on laboratory-prepared samples, or is for exposure conditions that are not exactly like those for the product under consideration. Thus, some characterization of the degradation behavior of the materials may be required to determine whether they will be able to meet all the design requirements over the expected life of the product.

3.4.5 **Ease of Use in Manufacturing Processes**

The materials must enable a product element to be easily produced. For a particular manufacturing process, some materials allow a product element to be more easily formed with consistent properties and features than other materials. Also, the properties of a material from one supplier may have less variation than a similar material from another supplier. Less variation in the properties makes it easier to use the manufacturing processes without problems. All these issues are discussed in more detail in Chapter 5.

3.4.6 **Total Cost to Use**

The actual cost to use a particular material includes the following expenses:

- The costs to purchase the materials used to create a product element.
- The costs to ship and store the materials. The properties of some materials degrade during storage before they are used to fabricate a component or build a subassembly or assembly. To prevent degradation, some materials must be packed in special packaging, shipped in a controlled environment, or stored in a controlled environment.
- The cost of the manufacturing processes required to form a product element from the selected materials.
- The cost to fix products with defective product elements before they leave a factory.
- The cost to throw away products that do not satisfy performance and reliability requirements.

The last three items fit in with the ease of manufacturing and are discussed in more detail in Chapter 5.

3.5 SELECTING THE MATERIALS

Based on the results of the evaluations, a design team will be able to determine which, if any, of the materials options will satisfy the requirements for a product element. If there is at least one acceptable material option for each product element, then the design team can continue on with the design. If there are not any acceptable material options for any of the product elements, then the design team must make one of the following choices:

Select a new design concept for the product element. This will involve identifying new design requirements. It may not be possible to change the design concept for one product element without selecting new design concepts for other product elements or selecting a new design concept for the subassembly or product. This depends on the impact of changing one product element on the rest of the design. Changing the design of a critical product element may force a complete redesign of the product.

Modify the product requirements. This will allow subassembly and product element requirements to be modified so that at least one material can be found that will satisfy the product element requirements.

Cancel the product. It may be determined that there is not enough time, money, or interest to select new design concepts, modify the product requirements, or invent a new material.

Invent a new material.

The first three options may not seem attractive. However, any one of them is better than spending the time and money on a development effort that results in a product that does not meet expectations. The fourth option requires an additional investment of time and money with no guarantee of success. However, if the design team is successful in inventing a solution, then the invention may give the company a competitive advantage over other companies.

The possibility of not being able to identify acceptable materials options is another reason why the constraints on the materials should be identified at the beginning of product development in order to plan and budget for any required materials testing and evaluation. By doing this, the development team can start to understand whether any of the product requirements must be modified while it is still early in the development cycle.

Material Properties and Materials Science

4

4.1 INTRODUCTION

This chapter explains the relationship between a material's properties, composition, and microscopic structures. Brief discussions of the materials science of different types of materials are provided. The discussions illustrate some of the many ways that composition and microscopic structures can be manipulated to achieve the desired properties. (Conversely, there are also many ways that the composition and microscopic structures, and resulting properties, can be compromised.) This chapter also provides information about available sources of materials science and engineering information.

Although it is not the purpose of this book to teach materials science, some degree of knowledge about it is required to understand the relationship between properties, composition, microscopic structures, and manufacturing processes, and to better appreciate the materials engineering perspective. The materials science information presented barely scratches the surface of each topic; however, what is here should be sufficient to provoke a healthy respect for the complexity of materials and the science related to them. The discussions should help dispel the myths that decisions about materials can be made intuitively and that common materials are not without their own intricacies.

4.2 MATERIAL PROPERTIES AND MATERIAL FEATURES

As discussed in the previous chapter, the performance and reliability of a product element depends on the properties of the materials from which it is made. The properties of any material depend on its various features—composition and microscopic structures.

The *composition* of a material refers to the atomic elements and compounds that make up a material. Different elements, mixtures of elements, and compounds have different properties. The intrinsic properties of any material depend on its composition. For example, tin melts at a lower temperature than copper, aluminum oxide is harder than magnesium oxide, and polycarbonate is stronger than polyethylene.

The *microscopic structure* is the manner in which the atoms, molecules, groups of atoms, and groups of molecules arrange themselves within a material. The sizes of microscopic structures cover a wide range, from nanometer scale to millimeter scale. The microscopic structure within a material depends on the following three factors:

1. The composition of the material
2. The manufacturing processes used to create the material and form it into a product element
3. The storage and use conditions to which a product element is exposed

The composition and certain microscopic structures can be manipulated as desired to achieve the desired properties of a material. This manipulation is achieved through the manufacturing processes used to create materials and form product elements. Specifically, the thermal, mechanical, electromagnetic, radiation, chemicals, and electrochemical conditions to which a material is exposed during manufacturing influence its microscopic structures. In addition, the use conditions to which a product element is exposed can affect a material's microscopic properties.

The diagram in Figure 4.1 illustrates the relationship between the following items: (1) the performance and reliability of a product, (2) the performance and

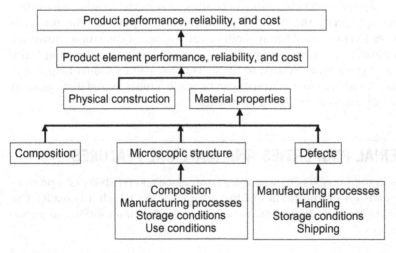

FIGURE 4.1

Factors underlying product performance and reliability.

reliability of product elements, (3) the physical construction and material properties of product elements, (4) the composition and microstructure of the materials composing each product element, (5) the defects within the materials, (6) the manufacturing processes, and (7) the use conditions.

The ability to select materials with properties that satisfy product element design requirements necessitates knowing the underlying composition and microscopic structures that influence the properties, as well as knowledge of how manufacturing processes and use conditions affect the composition and microscopic structures. In fact, understanding the relationship between composition, microscopic structures, materials properties, manufacturing processes, and use conditions is at the heart of materials science and engineering.

The ability to control the variation in composition and microscopic structures is crucial to controlling the variation of the properties of the materials in a product element. This is accomplished through proper control of the manufacturing processes used to form a product element and is discussed in more detail in Chapter 5.

The presence of manufacturing defects can cause the properties of a material to be degraded from the defect-free state, compromising the performance and reliability of a product element. *Defects* are unintended artifacts of the processes used to create a material and form it into its final shape. Some defects are difficult to prevent, but it is possible to minimize their size and number through proper manufacturing process practices. Others are a result of improper manufacturing process practices. Defects can also arise during handling, shipping, and storage of a component, subassembly, or product. Defects are internal to or at the surface of a material and can be microscopic or macroscopic in size depending on the type of defect and its severity.

The type, size, and number of defects that can be tolerated in a material depend on the particular performance and reliability requirements of the product element in which the material is used. So the presence of defects does not necessarily render a material unacceptable for use. The defects formed during various materials' preparation and manufacturing processes are discussed in Chapter 5.

4.3 CATEGORIES OF MATERIALS

Several broad categories of engineering materials (e.g., metals, ceramics, polymers, and composites) are used in products. Within each category, there are many subcategories of materials. In general, materials in each broad category have properties that distinguish them from other categories' materials. This is illustrated in Table 4.1, which shows the ranges of values for a variety of properties for metals, ceramics, and polymers.

New materials have also been developed that meet the definitions of metals or ceramics on a chemical basis; however, they have vastly different behaviors and properties compared to traditional materials. Such materials include nano-

Table 4.1 General Properties of Metals, Ceramics, and Polymers

Property	Metals	Ceramics	Polymers
Density, g/cm^3	2 to 22 (average 8)	2 to 19 (average 4)	1 to 2
Melting points	29.78°C (Gallium) to 3410°C (tungsten)	Up to 4000°C	Low
Hardness	Medium	High	Low
Tensile strength, MPa	Up to 2500	Up to 400	Up to 140
Compressive strength, MPa	Up to 2500	Up to 5000	Up to 350
Young's modulus, GPa	15 to 400	150 to 450	0.001 to 10
Thermal expansion	Medium to high	Low to medium	Very high
Thermal conductivity	Medium to high	Medium, but often decreases rapidly with temperature	Very low
Electrical characteristics	Conductors	Insulators	Insulators
Chemical resistance	Low to medium	Excellent	Good
Oxidation resistance	Generally poor	Oxides excellent; SiC and Si_3N_4 good	...

materials, quasicrystalline materials, and nonequilibrium amorphous metals and ceramics.

The next few sections briefly discuss the materials science of metals, ceramics, and polymers. Composites, surfaces, and interfaces are also discussed. The main goal here is to provide an appreciation for the microscopic structures that form within materials and the effects of those structures and material composition on the various properties of a material. Having this appreciation will help design and manufacturing engineers make better decisions related to material selection and control. Other materials (e.g., semiconductors, nanomaterials, quasicrystalline materials, and nonequilibrium amorphous metals) will not be discussed. The discussions of metals, ceramics, polymers, composites, coatings, surfaces, and interfaces should be enough to provide an appreciation for the importance of understanding the relationship between properties, composition, microscopic structure, and processing of materials.

4.4 METALS

A metal consists of a metallic element (e.g., copper, iron, nickel, gold) or is an alloy that has metallic properties and is composed of two or more elements, with the major constituent being a metallic element. On average, metals are good con-

Table 4.2 Examples of Metals and Their Applications

Metal	Applications
Steels	Motor shafts, fasteners, cans, automotive bodies, gears, tools, industrial equipment
Copper and copper alloys	Electrical contacts and wire, plumbing, coins, springs and clips, bearings, gears, kitchenware
Aluminum and aluminum alloys	Cans, automobile components, airplanes, appliances, bicycle frames, electrical conductors
Zinc and zinc alloys	Steel coating, automotive components, household appliance components, bearings, hand tools
Magnesium and magnesium alloys	Automotive, industrial, and aerospace die cast components; computer housings, hand tools, ladders
Nickel and nickel alloys	Gas turbine components, prosthetic devices, heat treating equipment components, rocket engine components, platings for electronics, ferromagnets
Titanium and titanium alloys	Airframe and jet engine components, prosthetic devices, components for chemical processing
Gold and gold alloys	Jewelry, conductors for electronics, platings for electronics and jewelry
Silver and silver alloys	Jewelry, platings for electronics, brazes and solders
Tin and tin-lead alloys	Solder, platings for electronics, coatings for steels

ductors of heat and electricity and have moderate to high strength. Some examples of metals and their applications are shown in Table 4.2. A small sample of textbooks and handbooks containing detailed information about specific metals, their applications, and properties can be found in the following references: Avedesian and Baker (1999); Brandes and Brook (1992); Chen (1986); Davis (1993, 1994, 2000, 2001); Donachie and Donachie (2002); Donachie (2000); Llewellyn (1994); McCurrie (1994); *Metals Handbook,* Volume 1 (1990); and *Metals Handbook,* Volume 2 (1990).

Metals are *crystalline* when in solid form. This means that the atoms within the metal are arranged in a pattern that is periodic, and the pattern continues in three dimensions within a single *crystal* of metal. A convenient representation of the periodic structure is shown in Figure 4.2 and is referred to as a *point lattice*. The places where the atoms reside are referred to as *lattice sites*. A point lattice consists of a repeating structure, referred to as a *unit cell*. In Figure 4.2(a), the unit cell is indicated by the light colored section. Figure 4.2(b) shows a more accurate representation of the unit cell structure. However, the model in Figure 4.2(a) is easier to use for demonstration.

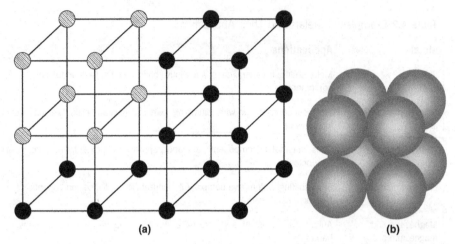

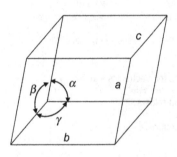

Crystal system	Edge lengths	Angles
Cubic	$a = b = c$	$\alpha = \beta = \gamma = 90°$
Tetragonal	$a = b \neq c$	$\alpha = \beta = \gamma = 90°$
Orthorhombic	$a \neq b \neq c$	$\alpha = \beta = \gamma = 90°$
Rhombohedral	$a = b = c$	$\alpha = \beta = \gamma \neq 90°$
Hexagonal	$a = b \neq c$	$\alpha = \beta = 90°; \gamma = 120°$
Monoclinic	$a \neq b \neq c$	$\alpha° = \beta = 90° \neq \gamma$
Triclinic	$a \neq b \neq c$	$\alpha \neq \beta \neq \gamma \neq 90°$

(c)

FIGURE 4.2

(a) Crystal lattice. (b) Solid ball model. (c) The seven crystal systems.

The size and shape of a unit cell is described by the lengths of the three axes that make up the unit cell and the angles between the axes. These lengths and angles are called the *lattice parameters* of the unit cell. A total of seven different kinds of unit cells are possible based on the various permutations of the lattice parameters. These different crystal systems are shown in Figure 4.2(c).

There are a total of 14 *lattice structures,* because it is possible to have atoms at positions other than just the corners. The 14 lattice structures are shown in Figure 4.3. For metals and alloys, the predominant lattice structures are face-centered cubic, body-centered cubic, and hexagonal. The actual arrangement of atoms in a crystal is referred to as the *crystal structure.* The crystal structure takes into account the different lattice structures, the lengths of the sides of the unit cell, and the angles between the axes. Although there are 14 possible lattice

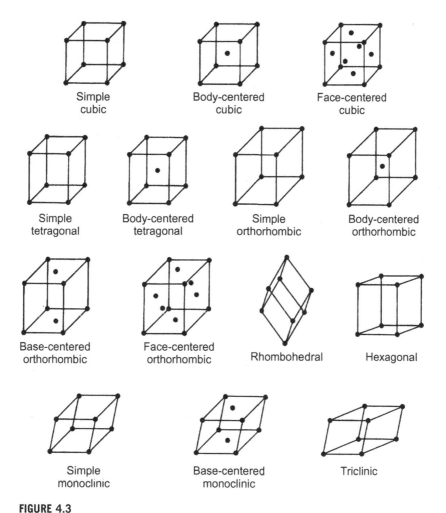

FIGURE 4.3

The 14 lattice structures.

structures, there is an infinite number of crystal structures when considering the different possible lattice parameters.

For any particular metal or alloy, the crystal structure is constant. However, because there are only 14 different lattice structures and thousands of different crystalline materials, many materials have the same lattice structure. For example gold, copper, nickel, and aluminum are face-centered cubic, whereas titanium, zinc, and magnesium are hexagonal. However, the lattice parameters are usually different from one material to another for the same lattice structure. For example, the lengths of the unit cells for gold, copper, and aluminum are 0.408, 0.361, and 0.405 nanometers, respectively.

A few different types of defects exist within the crystal lattice of a metal or alloy. These defects are different from the manufacturing artifacts mentioned earlier in this chapter. Vacancies, interstitials, and substitutions are three of them.

■ A *vacancy* is where an atom is missing from its lattice site. The number of vacancies in a metal crystal increases as the temperature increases.

■ An *interstitial* is an atom that occupies the space between the atoms on the lattice sites. Interstitial atoms must be small (e.g., hydrogen, carbon, and nitrogen) in order to fit into the spaces between the atoms on the lattice sites. The number of interstitial sites in a metal crystal is constant.

■ A *substitution* is the occupation of a lattice site by an atom of a different element.

Interstitials and substitutions arise from intentional and unintentional additions of an element to a metal. Schematics of a vacancy, interstitial, and substitution are shown in Figure 4.4.

The presence of vacancies and interstitials enables atoms to move from position to position through the crystal. This motion requires that a certain amount of energy in the form of heat be provided in order for atoms to move from one lattice position to another. The energy required for atom motion is different for different materials. The process of atom motion is called *diffusion,* and it enables various transformations to occur in a metal or alloy. These transformations result in changes in a material's properties. Understanding these transformations and their effects on a material's properties enables the control of the properties. Some examples of transformations will be provided later in this section.

Another type of defect in a metal's crystal lattice structure is a *dislocation,* which is an extra plane of atoms inserted into the crystal lattice. There are two types of dislocations: edge and screw. A schematic of an edge dislocation is shown in Figure 4.5(a). Dislocations move through the crystal lattice of a metal in response

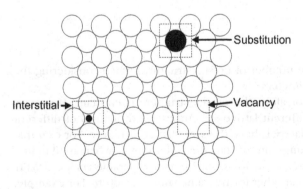

FIGURE 4.4

Schematic of an interstitial, vacancy, and substitution.

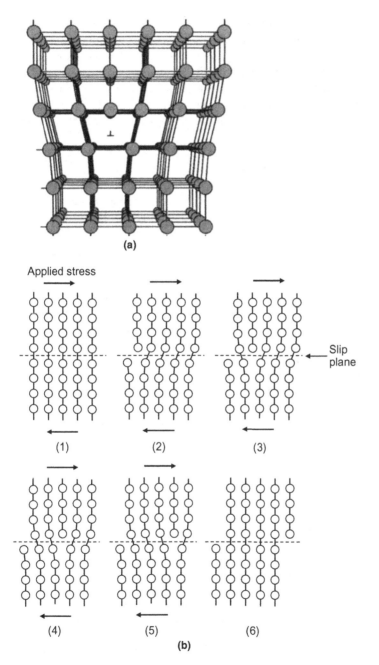

(a)

(b)

FIGURE 4.5

Schematics of (a) a dislocation in a crystal lattice and (b) a dislocation moving through a crystal lattice. (*Source:* Guy, 1976; part (a) reprinted with the permission of I. G. Guy.)

to applied stress, as shown in Figure 4.5(b). The motion of dislocations enables the deformation of a metal. For any particular metal or alloy, a minimum applied stress is required to cause dislocations to move. This minimum stress depends directly on the value of a metal's shear modulus. Hardened steel has a much higher shear modulus compared to aluminum. Therefore, it is more difficult to move dislocations in hardened steel than in aluminum. Dislocations are present in all crystalline metals, and the number and density of dislocations in a metal increases as the metal is deformed.

The arrangement of atoms with a metal crystal represents the smallest microscopic structure within a metal. The next largest structures within a metal are *grains* and *phases*. These structures will be explained in the next several paragraphs. Grains and phases within a metal can be observed using optical and electron microscopes at magnifications at least greater than about 100 times. Grain and phase structures that require magnifications greater than about 100 times in order to be observed are known as *microstructures*. Optical microscopes are useful for magnifications up to about 2000 times. High-magnification electron microscopes must be used to observe a metal's microstructure for grains and phases less than about 0.005 mm in size. Scanning electron and transmission electron microscopes can operate at magnifications up to about 100,000 and 1,000,000 times, respectively. Finally, metallic objects, such as castings, occasionally have very large grains that can be seen with the naked eye or with a low-power microscope. The structure in this case is called *macrostructure*.

Although very large single crystals of metal can be made, most metallic objects consist of many very small crystals. The individual crystals are referred to as *grains*, and the metallic object containing the grains is referred to as being *polycrystalline*. Figure 4.6 shows a picture of grains in a metal alloy. Within each grain,

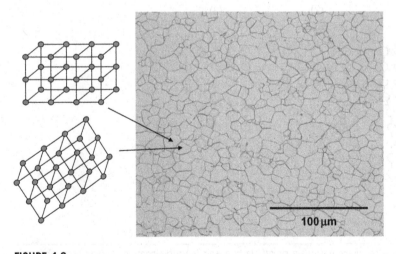

FIGURE 4.6

Micrograph showing the grains in a polycrystalline metal alloy.

the crystal lattice is continuous across the entire grain. The areas where two grains meet are referred to as *grain boundaries*. Grain boundaries are a result of the crystal lattices of the adjacent grains being rotated with respect to each other. This rotation occurs because the individual grains form independently, with the result that the orientation of the crystal lattice is different from grain to grain within a polycrystalline metal. Even though each grain within a metal has the same crystal structure, the lattices have a different orientation from grain to grain. The grain boundaries are narrow regions where the atoms try to arrange themselves to adjust for the differences in orientation between the lattices of neighboring grains. The result is that grain boundaries are regions of atomic disorder. A schematic of the arrangement of atoms in grain boundaries is shown in Figure 4.7. The schematic of the crystal lattice shown in Figure 4.6 demonstrates the possible orientation difference that occurs between adjacent grains.

In some alloys it is possible to minimize the extent of the differences in lattice orientation from grain to grain through special processing of the metal. These materials are referred to as being *textured*. Texturing gives a material enhanced mechanical properties compared to metals with grains that are randomly oriented from grain to grain.

Grains in a polycrystalline metal can range in size from a few nanometers to several millimeters in diameter, depending on the material and how it was

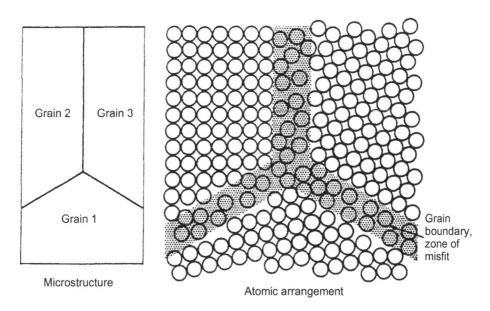

Microstructure   Atomic arrangement

FIGURE 4.7

Schematic of the arrangement of atoms in grain boundaries. (*Source:* Reprinted with the permission of ASM International.)

processed. Typically, the grain sizes within a metal sample are within a certain range that depends on the mechanical and thermal conditions to which the sample was exposed.

The grain size within a metal has an impact on various properties of a metal. For any particular composition metal, its strength increases as its grain size decreases. This occurs because the grain boundaries interfere with dislocation motion, and a small grain metal has more grain boundaries than a large grain metal. The electrical conductivity is lower for a metal with large grains compared to the same metal with very small grains. This occurs because the grain boundaries interfere with the flow of electrons through the metal.

A *phase* is a material with composition and properties that are homogeneous, and is physically distinct from other materials. One or more phases can be present within a metal sample. The number of phases present in a metal depends on the metal's composition and thermal history. Figure 4.8 shows a schematic of an alloy with two phases. The alloy consists of phases α and β. The entire alloy consists of elements A and B. However, the relative proportions of A and B in the α phase are different than the relative proportions of A and B in the β phase. That is, the compositions of the α and β phases are different.

Figure 4.9 shows an example of a microstructure in a nickel-vanadium alloy. The alloy consists of nickel with about 19 weight percent vanadium. Two phases are present. The dark-colored phase consists of a compound with the chemical formula Ni_3V—that is, nickel with 22.5 weight percent (25 atomic percent) vana-

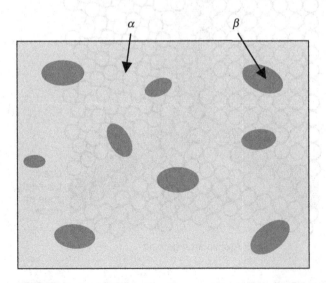

FIGURE 4.8

Schematic of metal with two phases.

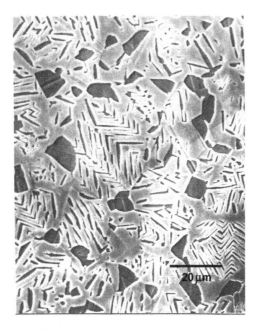

20 μm

FIGURE 4.9

Microstructure in a nickel-vanadium alloy.

dium. The lighter-colored matrix phase consists of a mixture of nickel and about 20 weight percent vanadium.

The *morphology* of grains and phases is also important. Morphology refers to the size and shape of the grains and phases present in a metal. The morphology of grains and phases is indicative of the composition and the mechanical and thermal treatment to which a metal has been exposed.

Many metal alloys contain different phases. The particular phases present in a material, their relative amounts, morphology, and spatial distribution depend on the composition of the alloy and the thermal and mechanical processes to which the alloy was exposed.

A phase can be (1) a solid solution or (2) an intermetallic compound. A *solid solution* consists of a solid phase in which one element acts as the solvent and the other as the solute, similar to a saltwater solution in which the water is the solvent and salt is the solute. However, in the case of metals, a solid solution is solid, not liquid. Solid solutions can be substitutional or interstitial. In a *substitutional solid solution,* the solute atoms occupy crystal lattice sites normally occupied by the solvent atoms. In an *interstitial solid solution,* the solute atoms reside in the spaces between the solvent atoms. A schematic showing both types of solid solution is shown in Figure 4.10. The white circles represent the solvent atoms, and the black circles represent the solute atoms. The light-colored matrix phase

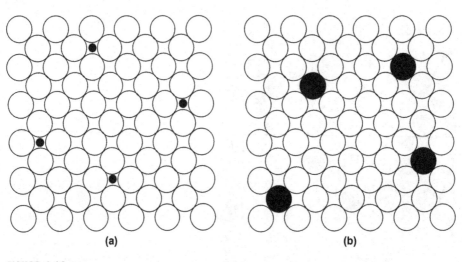

(a)   (b)

FIGURE 4.10

Schematic of solid solutions: (a) interstitial and (b) substitutional.

in Figure 4.9 is a substitutional solid solution with nickel being the solvent and vanadium being the solute. Many metal elements have the capacity to hold other elements in solution. For example copper can hold more than 30 weight percent zinc in solid solution at room temperature to make different brasses. The total amount of solute that can be held in solution depends on the specific solute element and solvent element and the temperature of the alloy.

Being able to form solid solutions gives engineers a "lever" to manipulate the properties of an alloy and enables the fabrication of alloys with important engineering uses. For example, some alloys can be strengthened and hardened by precipitation hardening. In these alloys, the amount of solute that can be held in solid solution decreases as the alloy temperature decreases. However, rapidly cooling the alloy from a high temperature can force the alloy to hold more than the equilibrium amount of solute in solid solution. In this state the alloy is referred to as a *supersaturated solid solution*. Then, controlled reheating of the alloy results in the formation of fine second-phase particles that strengthen and harden the alloy by impeding the motion of dislocations. This step is referred to as *aging*. The amount of strengthening and hardening depends on the pecific alloy composition, aging temperature, and aging time. These variables allow the properties of an alloy to be fine-tuned.

Intermetallic compounds are hard, brittle materials that form with simple whole-number fixed ratios of the component atoms. An example of an intermetallic compound is Ni_3V, which consists of three nickel atoms for every vanadium atom. Other examples of intermetallic compounds are Fe_3C, $AlCu$, Cu_5Sn_6, $NiAl$, and $AlSi$.

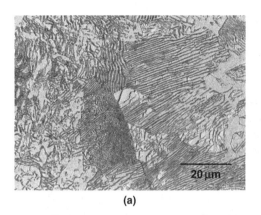
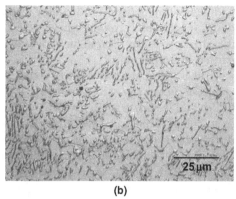

(a)   (b)

FIGURE 4.11

Microstructure of 1060 steel. (a) Lamellae of ferrite and iron carbide. (b) Globules of iron carbide in a ferrite matrix. (*Source:* Reprinted with the permission of Aston Metallurgical Services.)

The phases present in an alloy at room temperature, the relative amounts of the phases, and the morphology of the phases all depend on the alloy's composition, the temperatures to which it was exposed, the time of exposure, and the cooling rate from elevated temperatures. Furthermore, mechanical deformation, such as cold rolling, will influence the response of an alloy to thermal treatment. For an alloy of a particular composition it is possible to create different microstructures depending on the thermal and mechanical history of the material. Different microstructures will result in different properties.

An example of the effects of thermal processing on the phases present in an alloy is shown in Figure 4.11 for steel samples composed of iron with about 0.60 weight percent carbon and 0.70 weight percent manganese. The microstructure shown in Figure 4.11(a) is from a sample that was heated to 860°C and then air cooled. The microstructure consists of lamellae of the ferrite and iron carbide (Fe_3C) phases. The ferrite is light colored and the iron carbide is dark colored. The microstructure shown in Figure 4.11(b) consists of globules of iron carbide in a ferrite matrix. The globules of iron carbide are dark outlined. The microstructure was formed by heating the sample to 860°C, cooling to 650°C, and then held at 650°C to allow for the iron carbide to change to a partially spheroidal shape. The alloy with the lamellar microstructure is much harder and stronger than the alloy with the spheroidized microstructure. This is because the lamellar structure offers more resistance to dislocation motion compared to the globules of iron carbide.

A *phase diagram* is a valuable tool for understanding and predicting the phases and microstructures that can form within an alloy and for understanding the processes that were used to obtain a certain microstructure. The phases present within an alloy depend on the alloy's composition and temperature. *Alloy composition* is the total amount of each element in an alloy. Phase diagrams indicate the phases that can form in an alloy with a particular alloy composition and at a particular temperature.

A phase diagram for an alloy consisting of two elements has coordinates of composition (*x*-axis) and temperature (*y*-axis). As an example, the phase diagram for lead-tin is shown in Figure 4.12. The composition is given in weight percent or atomic percent. A phase diagram is divided up into *phase fields,* which indicate the phases that exist at any given composition and temperature. In a two-element alloy, the phase fields indicate a single phase or a mixture of two phases. This phase diagram consists of six phase fields indicated by (Pb), (βSn), Liquid, Liquid + (Pb), Liquid + (βSn), and (Pb) + (βSn). The regions marked (Pb), (βSn), and Liquid are all single phase fields. The (Pb) is solid lead with tin as a substitutional solid solute. The (βSn) is solid tin with lead dissolved as a substitutional solid solute. The Liquid is a mixture of liquid lead and tin. The other phase fields are mixtures of these three single phases—that is, Liquid + (Pb), Liquid + (βSn), and (Pb) + (βSn).

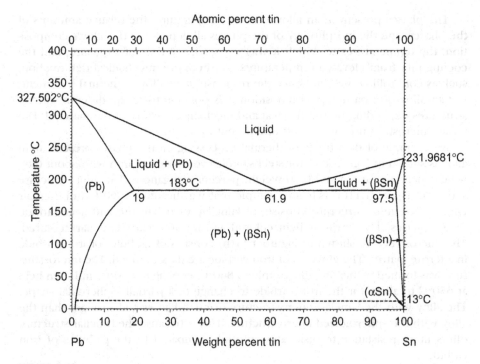

FIGURE 4.12

Lead-tin phase diagram. (*Source:* Humpston and Jacobson, 2004. Reprinted with permission of ASM International.)

At 25°C, the microstructure of any alloy between about 2 and 99% tin will contain a mixture of solid (Pb) and solid (βSn). The relative amount of each phase present depends on the specific alloy composition. As the alloy composition moves closer to the (Pb) region, the ratio of the amount of the (Pb) phase to the amount of the (βSn) phase present will increase. This is demonstrated in Figure 4.13, which shows the microstructure for samples with various alloy compositions. Figure 4.13(a) shows an example of the microstructure of a lead-tin with 70 weight percent tin. The dark-colored phase is the (βSn) phase, and the light-colored phase is (Pb). Figures 4.13(b) through (d) show the microstructure of lead-tin alloys containing 60% tin, 50% tin, and 40% tin.

The typical values for some of the properties of lead-tin alloys are given in Table 4.3. The values are based on measurements taken from samples cooled at

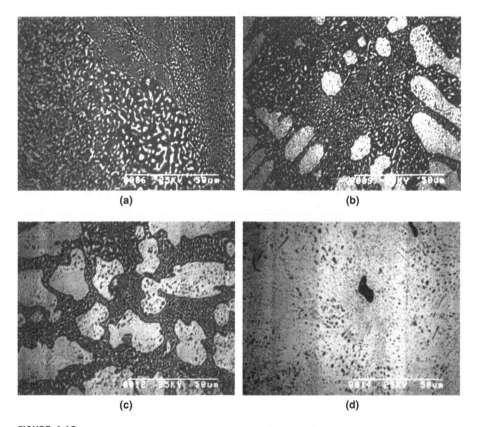

(a)   (b)

(c)   (d)

FIGURE 4.13

Microstructures in lead-tin alloys containing various amounts of tin: (a) 70% tin (micron bar), (b) 60% tin, (c) 50% tin, and (d) 40% tin. (*Source:* Hwang, 1996. Reprinted with the permission of McGraw-Hill.)

Table 4.3   Typical Properties for Lead-Tin Solders of Various Compositions

Tin content (weight %)	40	50	60	70
Density (g/cm^3)	9.28	8.90	8.52	8.17
Electrical resistivity ($\mu\Omega$-cm at 20 °C)	17.2	15.8	15.0	13.8
Thermal conductivity (W/m·K)	43.6	46.7	49.8	...
Coefficient of linear thermal expansion (10^{-6}/K)	24.7	23.6	21.6	20.7
Tensile strength (MPa)	37	41	52	54
Modulus of elasticity (GPa)	23	...	30	35

typical cooling rates from above the alloys' melting point temperatures. The data are reported as typical because the values will vary slightly because of variations in the cooling rate, which will give rise to variations in a solder's microstructure.

The phase diagram for lead-tin is an example of a relatively simple phase diagram. There are other two-element phase diagrams that are much more complicated. Figure 4.14 shows a diagram for the copper-zinc alloy system. This alloy system contains the following phases: copper solid solution (Cu), zinc solid solution (Zn), and the β, β', γ, δ, and ε solid solution phases, which form at different compositions of copper and zinc. Even more complicated phase diagrams can be constructed for alloys containing three and four elements.

The information provided by a phase diagram applies only to equilibrium conditions, which occurs when an alloy is held at a high temperature long enough to allow the atoms to diffuse to their preferred positions. However, many manufacturing processes do not allow for equilibrium to be reached. For example, a metal cooled rapidly from a high temperature to a low temperature may contain phases that are more characteristic of the higher temperature than they are of the lower temperature.

For many applications, nonequilibrium processing is used to obtain specific materials properties for an alloy. In these cases a phase diagram does not always provide enough information for correlating a microstructure to the thermal treatment of an alloy. This situation requires knowledge about the manner in which different phases form as a function of temperature, time, and the rates of heating and cooling, which is beyond the scope of the discussion in this chapter.

The thermal history of an alloy has an impact on both the phases formed and the morphology of the phases. Figure 4.15 shows the microstructures from three different samples of an aluminum alloy containing 5 weight percent silicon (Al-5 Si). The microstructures show three different types of particles within an aluminum matrix: silicon (dark gray), Fe_3SiAl_{12} (medium gray), and $Fe_2Si_2Al_9$ (light gray needles). Each of the samples was fabricated using a different casting process.

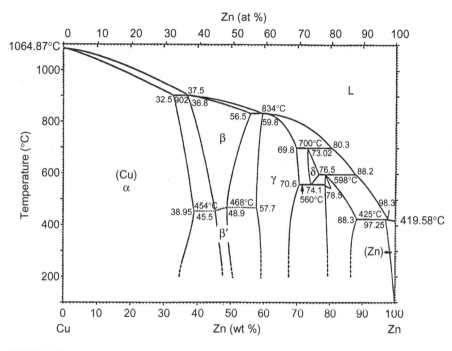

FIGURE 4.14

Copper-zinc phase diagram. (*Source: ASM Handbook,* Volume 3, 1993. Reprinted with the permission of ASM International.)

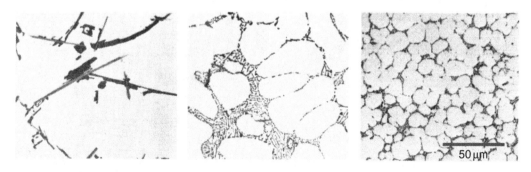

FIGURE 4.15

Al-5 Si microstructures resulting from different solidification rates characteristic of different casting processes. (*Source: ASM Handbook,* Volume 2, 1990. Reprinted with permission of ASM International.)

From left to right, the samples were sand cast, permanent mold cast, and die cast. The different casting processes resulted in three distinct microstructures because of the different solidification rates of the metal after the molten metal was poured into the molds. The solidification rate for die casting is fastest, followed by permanent mold casting, and then by sand casting. The size of the cells demarcated by the second-phase particles within the aluminum matrix can be seen to decrease as the solidification rate increased. Typical values of some of the mechanical properties for these materials are shown in Table 4.4. Notice that strengths increase as the process solidification rate increases and the cell size decreases.

Figure 4.16 shows the microstructures from sheets of a brass alloy composed of copper with 30 weight percent zinc that had been subjected to different defor-

Table 4.4 Typical Values of Mechanical Properties of Cast Al-5 Si

Process	Tensile Yield Strength (MPa)	Ultimate Tensile Strength (MPa)	Fatigue Properties
Sand cast	55	130	Good
Permanent mold cast	60	160	Good
Die cast	110	230	Excellent

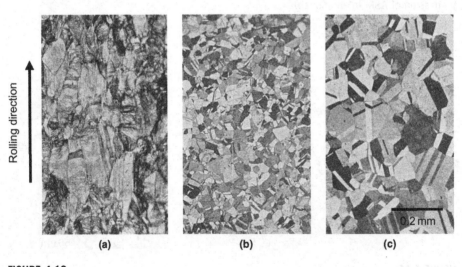

Rolling direction

(a)   (b)   (c)

0.2 mm

FIGURE 4.16

Cu – 30% Zn brass alloy (a) thinned to 50% of starting thickness by cold rolling; (b) processed as in (a) and then heated at 550°C for 1 hour; (c) processed as in (a) and then heated at 650°C for 1 hour.

mation and heating processes. The sample shown in Figure 4.16(a) was reduced in thickness by 50% from its original thickness by cold rolling. The grains are elongated and show a large degree of mechanical deformation. The sample shown in Figure 4.16(b) was cold rolled, just like the sample in Figure 4.16(a), then heated to 550°C for 1 hour. New grains are seen to have formed out of the as-rolled structure. The sample shown in Figure 4.16(c) was cold rolled just like the sample in Figure 4.16(a) then heat treated at 650°C for 1 hour. The higher heat-treating temperature allowed for the newly formed grains to grow larger compared to those in the sample in Figure 4.16(b). The average grain sizes for the samples in Figures 4.16(b) and (c) are 0.038 and 0.076 mm, respectively.

The strength and hardness of the as-rolled sample is higher than the sample in Figure 4.16(b), which is higher than the sample in Figure 4.16(c). The severe deformation of the material results in the rapid increase in dislocation density, which strengthens and hardens the alloy. The subsequent heat treatments resulted in annihilation of the dislocations and the formation and growth of grains with low dislocation density. The tensile yield strength and hardness for the three samples are shown in Table 4.5. Notice the dramatic decrease in yield strength and hardness after the heat treatment process compared to the as-cold rolled sample. Also, for the heat-treated samples, the hardness and strength decreases as grain size increases.

Table 4.6 shows the effects of composition on the electrical resistivity and thermal coefficient of resistance for copper-nickel alloys used as electrical resistors. Notice that the resistivity increases and the thermal coefficient of resistance decreases as the nickel content increases.

The magnetic properties of soft magnetic materials depend on their composition and microstructure. For nickel-iron soft magnetic alloys, three ranges of nickel content are commonly used: 36% nickel for maximum resistivity, 50% nickel for maximum saturation magnetization, and 80% nickel for highest initial and maximum permeabilities. A large number of elements can enter into solid solution, permitting tailoring of the magnetic and physical properties of the alloys. Many of these alloys are modified by adding molybdenum, silicon, manganese, copper, or

Table 4.5 Grain Size, Yield Strength, and Hardness for a Brass Alloy

Condition	Average Grain Size (mm)	Yield Strength (MPa)	Hardness (Rockwell B)
Cold rolled 50%	—	550	91
Cold rolled 50% and heated at 550°C for 1 hour	0.038	75	21
Cold rolled 50% and heated at 650°C for 1 hour	0.076	60	<10

Table 4.6 Variation of Electrical Resistivity and Thermal Coefficient of Resistance

Alloy	Resistivity (μΩ-cm)	Thermal Coefficient of Resistance (ppm/°C)
98Cu-2Ni	50	1400 (25-105°C)
94Cu-6Ni	100	700 (25-105°C)
89Cu-11Ni	150	450 (25-105°C)
78Cu-22Ni	300	180 (25-105°C)

chromium to produce specific magnetic properties. Heating nickel-iron alloys to high temperatures in a dry hydrogen atmosphere is performed to reduce their carbon, sulfur, and oxygen content and to allow for grain growth, all of which greatly improves the alloys' magnetic properties.

Finally, many different types of defects, such as voids, cracks, surface pits, and impurity particles, can be present in metal product elements. The kind of defects that can be created and their number depend on the specific manufacturing processes used to form a material into a product element. Some examples of defects in metals that can result from specific manufacturing processes will be provided in Chapter 5. The impact of a particular defect on the properties, performance, and reliability of a material depends on the specific material; the use conditions to which it is exposed; and the size, location, and number of the defect.

More information about the topics discussed in this section can be found in the following references: Barrett and Masalski (1980), Haasen (1996), Porter and Easterling (1989), Reed-Hill and Abbaschian (1992), and Verhoeven (1975).

4.5 CERAMICS

Ceramics are defined as inorganic, nonmetallic materials. This broad definition covers a wide range of materials that extend far beyond the common notions of ceramics and glasses. Ceramics are used across a broad range of structural, electronic, thermal, optical, nuclear, biomedical, and chemical processing applications. In fact, certain families of ceramics have desirable optical, electrical, and magnetic properties that can be tailored to specifications based on specific atomic bonding and crystal structures.

In general, ceramic materials are hard, brittle, wear resistant, and corrosion resistant. These properties reflect the strong atomic bonds that exist between atoms in many ceramic materials.

From a chemical composition perspective, ceramics are commonly chemical compounds rather than single element compositions. These ceramic compounds, which are two or more elements in combination, are commonly organized into groups of metal-based oxides, carbides, nitrides, and borides. Metals commonly found in ceramics include iron, aluminum, titanium, zirconium, magnesium, molybdenum, silicon, and calcium. Other materials, such as silicon carbide, gallium arsenide, magnesium fluoride, molybdenum disilicide, and diamond (an elemental carbon), are also considered to be ceramics even though they are outside the conventional composition groups.

Ceramics include crystalline materials and noncrystalline amorphous glasses, which are commonly found as oxide compositions. In glassy/amorphous materials, there is not a periodic arrangement of atoms for distances greater then a few atoms in any direction from any particular atom. This is referred to as *lacking long-range order*.

The discussion in this section focuses on crystalline ceramics. Information about glasses is available in Doremus (1994), *Engineered Materials Handbook* (1991), and Shelby (2005).

4.5.1 Applications

From an applications standpoint, crystalline and glass ceramics are divided into two categories: traditional and advanced.

Traditional Ceramics

Traditional ceramics are divided into the following categories:

1. Structural clay products such as bricks, tiles, and pipes. These products consist of various combinations of mostly silica and alumina, with smaller amounts of other oxides such as iron oxide, magnesia, titania, potassium oxide, and sodium oxide.

2. Whitewares such as stoneware (tableware, artware, tiles, and cookware), china (tableware, artware, cookware, toilets), porcelain (tableware, artware, electrical insulators, dentures), and electrical insulators. These products use the same compounds found in structural clay products.

3. Cements, such as concrete and mortars, are synthetic mineral mixtures that are used in the construction of roads, bridges, and buildings. Cements form a hard, bonded mass through an ambient temperature hydration reaction, which does not require heat. The four main compounds in cement are tricalcium silicate, dicalcium silicate, tricalcium aluminate, and tetracalcium aluminoferrite.

4. Refractory materials are used as thermal insulation in high-temperature furnaces. They resist degradation by corrosive gases, liquids, or solids at elevated temperatures. Examples of refractory materials are silica, aluminum silicate, and magnesite.

5. Traditional glass applications include mirrors, windows, containers, lighting fixtures, and glass fibers for thermal insulation and composites. Almost all

glasses are based on silica with the addition of other oxides—boria, calcia, alumina, and other materials.

Traditional ceramics and glasses are commonly oxide compositions and are produced in both crystalline and amphorous states. In many forms they are not fully dense and contain porosity at the micron and higher size range. In some applications (e.g., thermal insulation, filters) the porosity may have a distinct engineering function such as controlling thermal properties and determining the mechanical strength.

An example of the microstructure of a high-temperature 50% alumina, alumina-silica refractory ceramic (used for thermal insulation) is shown in Figure 4.17. Products of this type may be used in a variety of less demanding applications such as carbon baking furnaces, incinerators, and ladle backup linings. Figure 4.17(a)

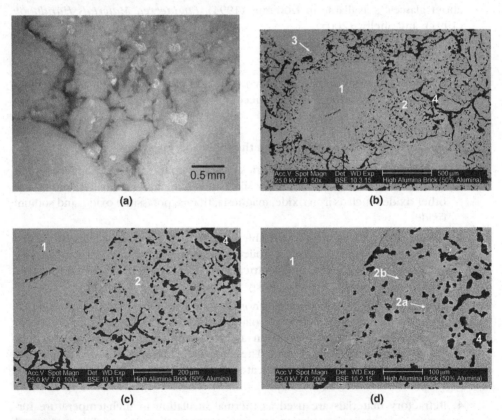

(a)

(b)

(c)

(d)

FIGURE 4.17

Microstructure of a 50% alumina, alumina-silicate refractory ceramic. (a) Optical image showing the alumina-silica grains; (b) through (d) scans of electron images showing the fine microstructure within the sample. (*Source:* Reprinted with the permission of ANH Refractories Company.)

is an optical image of the material, which shows coarse and intermediate alumina-silica grains and voids. Figures 4.17(b-d) are scanning electron images of the same material at different magnifications. The alumina-silica grains are designated by the 1 in Figures 4.17(b-d). The matrix, designated by the 2 in Figures 4.17(b) and (c), consists of the mullite and cristobalite phases bonded with a glass phase. Mullite is a compound consisting of alumina (aluminum oxide) and silica (silicon dioxide). Cristobalite is a form of silica (silicon dioxide). The mullite/cristobalite (white) and glass (gray) phases are clearly seen in Figure 4-17(d) and are designated by 2a and 2b, respectively. Also present are iron oxide and titanium oxide accessory phases (3 in Figure 4.17(b)) and voids and pores (4 in Figure 4.17(b) and (c)).

The images in Figure 4.17 show that, just like metals, multiple phases can be present in ceramic materials. As with metals, the phases present, their relative amounts, and their morphology all depend on the composition of the material and its thermal history.

Advanced Ceramics and Glasses

Advanced ceramics and glasses are carefully engineered materials that have been developed for their particular mechanical, electrical, magnetic, or optical properties. Advanced ceramics and glasses include a wide range of chemical compositions, crystal phases, and varying microstructures. They can be classified in the following application areas:

- Structural ceramics for high-stress, high-temperature, or corrosive environments where metals fail. Applications include engines, wear parts, thermal management, processing equipment, and biomedical and dental components. Compositions in this category include alumina, silicon nitride, silicon carbide, zirconia, boron nitride, mullite, cordierite, and molybdenum disilicide.

- Electronic ceramics for capacitor, sensor, semiconductor, superconductor, ionic conductor, varistor, and substrate/packaging applications. Alumina, aluminum nitride, and boron nitride are used for electronic substrates. Titanate compositions are used for capacitors, sensors, actuators, and ferroelectrics.

- Optical ceramics and glasses for information display, solid-state lasers, optical data transmission, photoelectrics, wavelength-tuned windows, and engineered mirrors. Common composition families are titanates, rare-earth oxides, and chalcogenide glasses.

- Wear and corrosion resistance for pump components (seals, bearings, plungers), faucets (seals, valves), and rotary equipment (bearings). Common compositions include alumina, zirconia, silicon carbide, and silicon nitride.

- Magnetic ceramics for actuator, sensor, information storage, transformer, recording, sonar, and microwave applications. These include ferrites, rare earth oxides, and copper oxide compositions.

- Machining and grinding tools for long life and high efficiency, using diamond, silicon carbide, tungsten carbide, silicon nitride, and alumina.

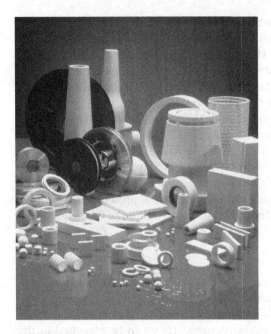

FIGURE 4.18

Examples of ceramic components. (*Source:* Reprinted with the permission of CoorsTek, Inc.)

■ Ceramic coatings for wear, abrasion, and corrosion resistance, based on titanium carbide and nitride, diamond compositions, silicon carbide, and others.

Examples of ceramic components are shown in Figure 4.18.

4.5.2 Atomic Bonding and Crystal Structure in Ceramics

The bonding structure and crystal structure within a ceramic material are fundamental physical features that determine the intrinsic physical, mechanical, thermal, electronic, and magnetic properties of a given ceramic. The atomic bonding and crystal structure in ceramics are determined by the chemical composition.

The atoms in a ceramic are bonded together with one of two types of chemical bonds, ionic or covalent. The two types of bonds result in materials with different properties. Ceramics with covalent bonds have higher electrical conductivity, higher melting points, and higher strength compared to ceramics with ionic bonds. However, bonding in ceramics is not simply ionic or covalent. In actuality, the bonding in ceramics crosses a spectrum from ionic to covalent, with different degrees of the extent of ionic and covalent character from material to material; for example, MgO is 73% ionic. The properties of a ceramic depend to some extent on the bonding character within the materials.

The atomic bond strength in a ceramic is the primary determinant of the melting temperature, the elastic modulus, and the inherent mechanical strength.

This is because atomic bonds are stretched and broken as a material melts or is mechanically strained. Ceramics with strong atomic bonding melt at high temperatures and are very stiff and hard. Ceramics with weak atomic bonding have low melting temperatures, low elastic modulus, and are relatively soft.

The crystal structures for ceramics are commonly more complicated than for metals, because ceramics are found as compounds that consist of two, three, or four elements. Various crystal structures are found that accommodate different bonding structures, compositions, and sizes of atoms. In contrast to metals, where the crystal structure nomenclature is defined by the crystalline structure, ceramic crystal structures are named after the compound for which the structure was first characterized. The following are a few examples of ceramic crystal structures:

- Rock salt structure, named after NaCl
- Cesium chloride
- Zinc blend (ZnS)
- Perovskite. Compounds exhibiting the perovskite crystal structure include $CaTiO_3$, $BaTiO_3$, $SrTiO_3$, $PbTiO_3$, $PbZrO_3$, $LaBa_2Cu_3O_{7-x}$, and other oxide superconductors.

These are shown in Figure 4.19. Other ceramic crystal structures include rutile, wurtzite, corundum, and spinel.

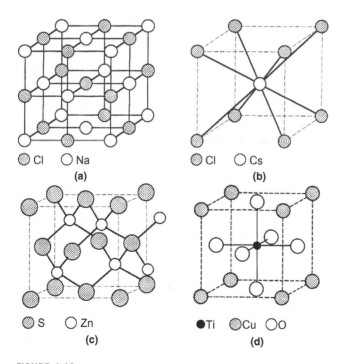

FIGURE 4.19

Rock salt (NaCl), cesium chloride, zinc blende, and perovskite crystal structures.

The thermal properties of a ceramic are determined in large part by its atomic composition and crystal structure. Unlike metals, ceramics seldom have electron bonding with free electrons. As a result, thermal conduction in ceramics occurs through vibration of the bonds between atoms and radiation between atoms.

The type of atomic bonding in ceramics is the primary determinant of the electronic conductivity in them. The following are the five different classes of electric conduction for ceramics:

1. *Insulators.* These are poor conductors of electricity. Most ceramics have neither mobile ions nor electrons at temperatures below a few hundred degrees Celsius. They act as electrical insulators and include silica, alumina, silicon nitride, and magnesia. However, impurities can have a dramatic effect on the conductivity of these materials, with the electrical conductivity increasing as the concentration of impurities increases.

2. *Ionic conductors.* In ceramics with strong ionic bonding, electrical charge is carried by ions. If the ions are sufficiently mobile and the quantity of them is large enough, these ceramics can conduct appreciable electricity as a result of ions moving through the crystal lattice. Such ceramics are referred to as *fast-ion* ceramics and include zirconia, beta-alumina, and alkali halides.

3. *Semiconductors.* Ceramic semiconductors can conduct electricity when electrons are energized into a conductive state. The source of energy can be thermal, radiational, and electromagnetic. There is wide range of such semiconductor ceramics, including iron oxide, silicon carbide, copper oxide, zinc oxide, and titanates.

4. *Electron conductors.* Some ceramics conduct electricity the same as metals with free electrons, but these are relatively rare. These materials are transition metal oxides such as rhenium oxide, chromium oxide, and titanium monoxide.

5. *Superconductors.* In almost all materials, heat is generated as electrons flow through the material as a result of electrical resistance. However, in superconductors, there is no electrical resistance and electrons flow unimpeded. Ceramic superconductors based on yttrium-barium-copper oxide compositions exhibit superconductivity at temperatures of 90 K, which is significantly higher than the temperature (<23 K) that is needed for conventional metallic superconductors.

In both ionic and semiconductor ceramics, impurities and crystal defects play key roles in determining tight electrical conductivity. Impurities introduce electrical charge imbalances, which provide a source of electrical charge carriers. Crystal defects (i.e., vacancies, etc.) provide open sites for ions to move through the lattice. Both types of defects enable increased electrical conductivity. For example, at elevated temperatures, oxygen ions in doped zirconium oxide can move through the lattice from vacancy to vacancy when an electrical field is applied.

4.5.3 Microstructure Features

Many different types of microstructure features need to be considered in both polycrystalline and glass ceramics. These features include grain structure; phases and phase morphology; and defects such as pores, inclusions, cracks, and scratches. As with metals, the microstructure present in a ceramic depends on its composition and the manufacturing process conditions to which it has been exposed.

For most ceramic components, the manufacturing process consists of *sintering* a component made of a loosely joined ceramic powder, which has been formed into the component shape by processes such as pressing, tape casting, and slip casting. Sintering involves heating the formed material to a high enough temperature for the ceramic powder particles to fuse together.

The polycrystalline grain structure within a ceramic material is defined by grain size, shape, and orientation. Grain size and structure depends on the compositions and the phases that are present and on how a ceramic is thermally processed. Ceramics that have a finer grain size generally have higher strength but lower high-temperature creep resistance.

Ceramics may contain multiple phases just like metal alloys. Figure 4.20(a) shows an example of the microstructure for a 97.5% alumina (Al_2O_3) that has a glassy silica (SiO_2) phase at the grain boundaries. Figure 4.20(b) shows the microstructure for a 99.5% alumina, which has much less silica. The 99.5% alumina is stronger and harder than the 97.5% alumina. Typical data for these properties are shown in Table 4.7. Secondary phases at the grain boundaries commonly arise from solidification of phases that turn to liquid during sintering. Secondary phases can be glassy (amorphous) or crystalline depending on their composition. Notice that the properties' values are lower for the alumina with more glassy

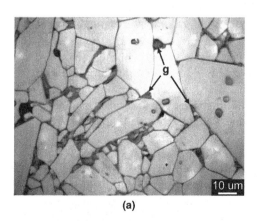
(a)

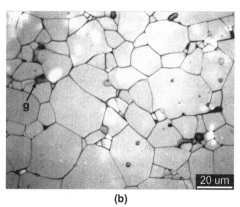
(b)

FIGURE 4.20

Microstructure of alumina samples. (a) 97.5% alumina; (b) 99.5% alumina. (*Source:* Reprinted with the permission of Carpenter Ceramics.)

Table 4.7 Typical Properties for 97.5 and 99.5% Alumina

	97.5% Alumina	99.5% Alumina
Flexural strength (MPa)	292	344
Modulus of elasticity (GPa)	345	374
Hardness (Rockwell 45N)	76	85

silica at the grain boundaries. Secondary phases can reduce the strength, thermal stability, electrical conductivity, and thermal conductivity of a ceramic.

The presence of a secondary phase located on the grain boundaries of the primary phase can have significant effects on the mechanical, thermal, and electronic properties of a ceramic. The purity of the powders used to fabricate dense ceramic components determines how much of a grain boundary phase develops. If an insoluble phase is formed during component fabrication, that phase may segregate to the grain boundaries. For example, a high level of silica and calcia impurities in alumina leads to the formation of a glassy phase at the grain boundaries, as illustrated in Figure 4.20. This glass phase has lower strength and a reduced melting temperature, and it will reduce the component's high-temperature strength and creep resistance compared to a more pure material.

The pores found in ceramics are commonly an artifact of the manufacturing process used to produce a component. These pores can exist in a range of sizes, geometries, and locations. The pores in ceramics can range from nanometers to centimeters in diameter. The size and distribution of pores in a ceramic component are determined by the raw materials and processes used to make the component. Figure 4.21 shows the pores in an alumina sample (Coble & Burke, 1963). Pores can be isolated within a solid or in the form of interconnected channels that thread through the solid to the surface. These are referred to as *closed porosity* and *open porosity*, respectively. Pores can reside between and within grains.

The total pore structure influences the properties of a ceramic component. For example, the direct current electrical conductivity and the thermal conductivity decrease as the volume fraction of pores increases. In a similar manner, mechanical strength, elastic modulus, and hardness all decrease with higher levels of and larger sized pores.

Inclusions are undesired particles that remain in the ceramic or glass as an artifact of manufacturing. The presence of inclusions will affect the grain size of a ceramic material because the inclusions interfere with grain growth during sintering. If there is 1% by volume of inclusions, the ceramic grains can only grow to approximately 100 times the inclusion size. For 10% by volume of inclusion, the grains can grow to only 10 times the inclusion size. The effectiveness of inclusions in limiting grain growth increases as inclusion size decreases and its volume fraction increases. Cracks are effectively sharp edge pores that act as stress

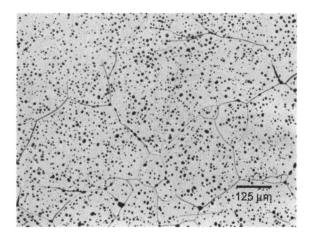

FIGURE 4.21

Porosity in alumina. (*Source:* Coble and Burke, 1963.)

concentrations. Scratches are surface cracks, commonly produced by mechanical contact with harder materials. Inclusions, cracks, and scratches commonly reduce the mechanical strength of ceramics and glass.

4.5.4 Phase Diagrams

As with metals, phase diagrams are used to describe the phases that are present in a ceramic material system as a function of composition and temperature. In the case of ceramics, a phase diagram is defined by combinations of compounds rather than elements. Figure 4.22 shows a phase diagram for SiO_2 and Al_2O_3 (Aramaki & Roy, 1962).

Different phases in a ceramic have different mechanical, thermal, electrical, and magnetic properties. The phases can be modified to provide the desired performance for a given application. For example, zirconia-toughened alumina contains a dispersed zirconia phase within an alumina matrix. The presence of the zirconia gives a higher mechanical strength and fracture toughness, compared to pure alumina.

4.5.5 Ceramic Strength

Unlike metals, the dislocations in most ceramics do not offer an effective mechanism for plastic deformation. Therefore, cracks and flaws act as stress concentrators in ceramics, lead to failure in a brittle manner when stresses and strain energy reach a level where the cracks and flaws propagate catastrophically. Thus, the brittleness of ceramic materials is caused by their susceptibility to unstable crack growth. Large cracks will result in fracture at lower stresses than small cracks. Cracks in ceramics originate at small internal defects, such as pores or inclusions,

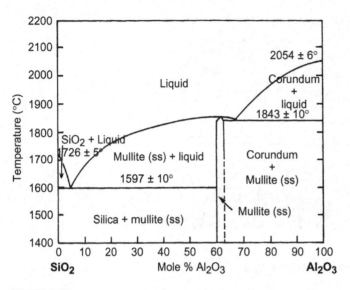

FIGURE 4.22

Phase diagram for SiO_2 and Al_2O_3. (*Source:* Reprinted with the permission of Blackwell Publishing.)

or are initiated because of mechanical damage during processing and handling. A large grain in a ceramic may also act as a critical flaw.

The fracture toughness of a ceramic is a material property that describes the resistance of a material to crack propagation. The relationship between the fracture strength σ_s of a material and the size of a flaw or crack c is given by

$$\sigma_s = K_{IC}/(\pi c)^{1/2}$$

where K_{IC} is the fracture toughness. The fracture strength decreases as the crack size increases. The fracture strength of a ceramic with a certain fracture toughness depends on the size and location of the largest crack/flaw in the ceramic. There is large variation in the strength from sample to sample for particular component because of the variation of flaw sizes in the different ceramics.

The fracture toughness and engineering strength of ceramics can be improved by engineering the microstructure so that higher energy is needed for a crack to grow. One example of this is in partially stabilized zirconia. Zirconia can have either a cubic crystal lattice structure or a tetragonal crystal lattice structure. The fracture toughness of cubic zirconia is about 2 MPa·m$^{1/2}$. Adding a small amount of magnesia to zirconia and using a special thermal treatment results in the creation of partially stabilized zirconia, which has a fracture toughness up to five times that of cubic zirconia. This material consists of grains of zirconia with a cubic crystal lattice, and submicron second-phase zirconia particles that have a tetragonal crystal lattice structure. Micrographs of partially stabilized zirconia are shown in Figure 4.23. In (a), only the large second-phase particles at the grain

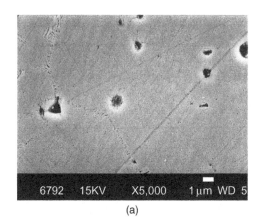
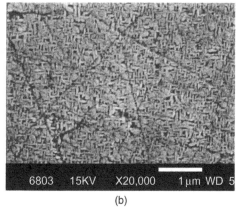

(a)   (b)

FIGURE 4.23

(a) Partially stabilized zirconia. (b) Higher-magnification image. (*Source:* Reprinted with the permission of Carpenter Ceramics.)

boundaries can be seen. In (b), the higher magnification image enables the smaller particles inside of the grains to be seen.

4.5.6 More Information

Additional detailed information about ceramic materials, their applications, and ceramic manufacturing processes can be found in Barsoum (1997); *Engineered Materials Handbook,* Volume 4 (1991); Kingery, Bowen, and Uhlmann (1976); Lee and Rainforth (1994); Richerson (2006); and Van Vlack (1980).

4.6 POLYMERS

A *polymer* is a long molecule that consists of a chain of many units of a smaller molecule joined together by chemical bonds. The smaller molecule is referred to as a *mer unit*. Figure 4.24 shows a schematic that demonstrates how individual ethylene mer units are joined together to form a polyethylene molecule (Lampman, 2003). Figure 4.25 shows some examples of the mer chemical structure for different polymers. Usually, the polymer molecule chains are long, often consisting of hundreds or thousands of mer units. However, polymers consisting of only a few mer units joined together are available.

A polymer is produced through a process called polymerization whereby mer units are reacted together chemically to form large molecules. In all polymers, the polymerization process results in polymer chains of various length within the material. In some polymerization processes, very long chains are formed. In other processes, the chains are considerably shorter. All polymerization processes result in polymer chains of various length within a material. The length of polymer

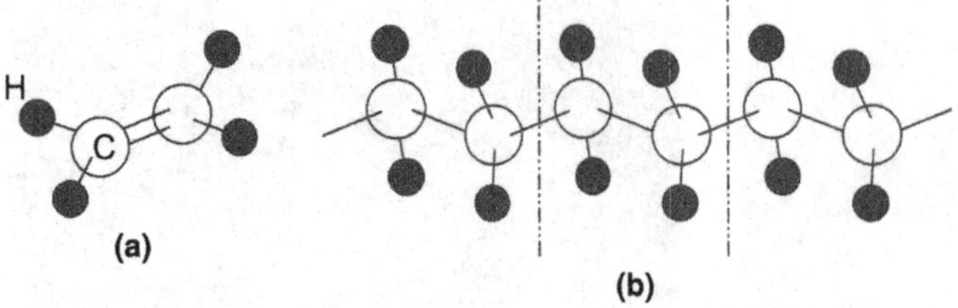

FIGURE 4.24

Schematic of the (a) ethylene mer unit and the (b) polyethylene molecule. (*Source:* Lampman, 2003. Reprinted with the permission of ASM International.)

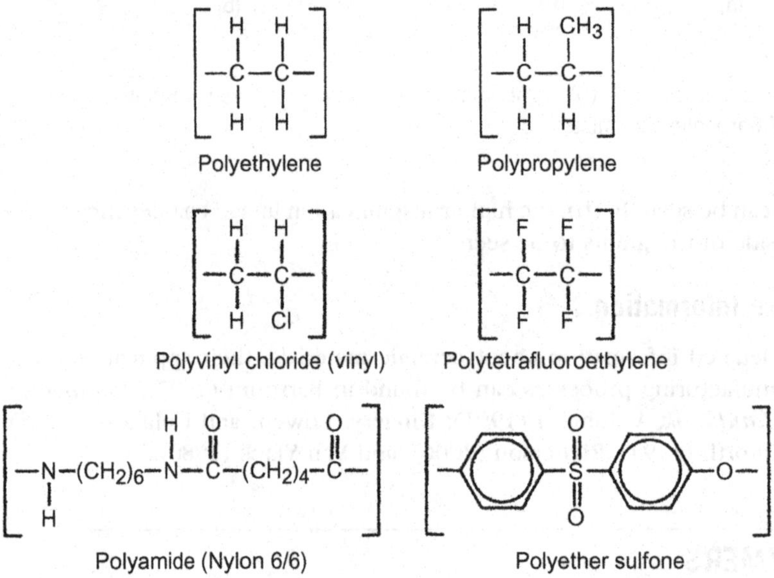

FIGURE 4.25

Mer chemical structure for various polymers.

chains in a material is measured by the average molecular weight and the molecular weight distribution. The *average molecular weight* is the average weight of the polymer chains, which is directly related to the lengths of the polymer chains in a material. As the chain length increases the molecular weight increases. Furthermore, because there is a range of chain lengths within a sample, there will be a *molecular weight distribution*. It is possible to make polymer samples that have the same molecular weight, but with different molecular weight distributions. The molecular weight distribution for a particular polymer depends on the specific process used to polymerize the material.

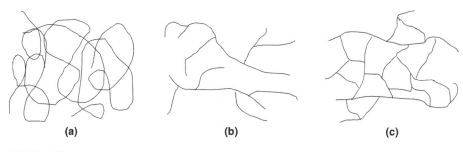

FIGURE 4.26

Arrangements of polymer molecules: (a) Linear. (b) Branched. (c) Three-dimensional
network (i.e., cross-linked).

Polymers are typically low density, not electrically conductive, and have
maximum use temperatures up to about 250°C. Some polymers are naturally occurring and others are human-made. Some examples of naturally occurring polymers
are cotton, wool, silk, wood, starch, and protein. Only the synthetic polymers, such
as plastics, elastomers, and gels, are of interest for this book.

Polymer molecules can be linear, branched, or in the form of a three-dimensional
network, as shown in Figure 4.26. Linear molecules are just long polymer chains.
Branched molecules consist of chains with branches that are chemically bonded to
the main chain. Networked molecules have chemical bonds between chains in addition to the chemical bonds between the mer units that form the main polymer chain.
This bonding between chains is referred to as cross-linking.

4.6.1 Types of Polymers

Polymers can be categorized into the following groups: thermoplastics, thermosets, and elastomers.

Thermoplastics consist of linear or branched molecules. These materials soften
and flow when heated. The temperature at which softening first begins is called
the *glass transition temperature*. The temperature at which thermoplastics
melt is called the *melting point temperature*. Theoretically, thermoplastic
materials can be repeatedly reheated and reshaped.

Thermosets consist of a dense, three-dimensional molecular network with a large
degree of cross-linking between polymer chains. The result of the cross-linking
is a rigid material. Thermosets do have a glass transition temperature; however,
they do not have a melting temperature. Instead of melting they char and
decompose when exposed to high temperatures. Once reacted and formed,
thermosets cannot be reformed by heating.

Elastomers are either thermoplastics or thermosets that have a lightly linked
molecular network, and are sometimes placed in their own category. The

lightly linked molecular network allows the molecules to slide past each other as an elastomer is deformed but the cross-links prevent permanent flow. As a result, elastomers can be easily stretched to high elongations and will return to their original shape when the stress is removed.

4.6.2 Polymer Composition and Microscopic Structure

Some examples of thermoplastics, thermosets, and elastomers and their applications are shown in Table 4.8.

The properties of a polymer are dictated by its composition and structure. The polymer structure is described by the composition and structure of the mer unit, the structure within the polymer molecule, and the structure between polymer molecules.

Mer Unit

The atoms within a particular type of mer unit are joined together in a specific arrangement that is characteristic of the molecule. The elements involved, their arrangement with respect to each other, and the strength of the chemical bonds between the atoms determine the chemical and physical properties of a mer unit. The chemical properties of a molecule determine the types of reactions into which the molecule can enter. For instance, not all molecules can be combined into long chains to form a polymer.

A mer unit consists of a backbone and side groups. The backbone consists of the atoms that lie along the length of a polymer chain. For ethylene, the two carbon atoms make up the backbone. The side groups consist of atoms or groups of atoms attached to the backbone. For ethylene, the hydrogen atoms make up the side groups.

Table 4.8 Selected Human-Made Polymers and Their Uses

Polymer	Class	Uses
Polyethylene (PE)	Thermoplastic	Plastic bags, tubing, electrical insulation
Polyvinyl chloride (PVC)	Thermoplastic	Plastic pipe, rain gear
Polytetrafluoroethylene (PTFE)	Thermoplastic	Nonstick coatings for cookware, dental floss, low-friction coatings for industrial equipment
Polyvinyl acetate (PVA)	Thermoplastic	Chewing gum, adhesives
Nylon	Thermoplastic	Electrical connectors
Epoxy	Thermoset	Adhesives, printed circuit boards, matrix for composite materials
Polyisoprene (rubber)	Elastomer	Hoses, tires, gloves

Table 4.9 Bond Energies for Common Bonds in Polymers

Bond	Bond Energy (kcal/gram·mole)
C–C	83
C–H	99
C–F	105
C–Cl	79
C–O	84
N–N	38
C=C	194
C=O	171

Note: – indicates single bond; = indicates double bond.

The flexibility and bulkiness of the mer unit will affect the interactions between polymer chains and thus the arrangement of the polymer chains within a material. The flexibility of the mer unit decreases as the chemical bond strengths between the backbone atoms increase. Table 4.9 shows the bond energies for common bonds in polymers. The bond strength depends on the elements bonded together and the type of bond that exists between atoms. Certain groups of backbone atoms, such as aromatic rings or cyclic groups, will also reduce flexibility and add bulkiness to the mer unit. Polyether sulfone is an example of an aromatic ring containing mer units (see Figure 4.25). The bulkiness of the mer unit is also increased by the addition of large, inflexible side groups to the mer.

The flexibility and bulkiness of the mer unit has a variety of effects on the polymer. For example, the melting point temperature of thermoplastics increases as chain flexibility decreases, chain bulkiness increases, and as intermolecular bonding strength increases. Also, as the backbone stiffness of a polymer increases, strength, impact toughness, and thermal properties will generally increase.

The chemical nature of thermoplastics, thermosets, and elastomers is based on the elements present in the mer unit and how they are arranged. Thermoplastics have four categories of polymer materials—hydrocarbon, carbon-chain, hetero-chain, and aromatic ring-containing polymers—based on the mer unit's chemistry.

Hydrocarbons are low molecular weight polymers that consist of hydrogen and carbon. Examples of hydrocarbons are polyethylene and polypropylene, both shown previously in Figure 4.25.

Carbon-chain polymers have chlorine, fluorine, oxygen, and nitrogen in the side groups, such as polyvinyl chloride (PVC) and polytetrafluoroethylene (PTFE).

Heterochain polymers are often stronger and have higher temperature resistance than carbon-chain polymers. In addition to carbon, the backbone of heterochain polymers contains at least one other element such as oxygen, nitrogen, sulfur, and silicon. An example is polyamide (nylon 6/6), shown in Figure 4.25.

Aromatic ring containing polymers are a subset of heterochain polymers and contain the benzene ring in the backbone. An example of an aromatic ring containing polymer is polyether sulfone, which is shown in Figure 4.25. These materials have very rigid chains giving them excellent high-temperature mechanical properties. Most high-temperature thermoplastics have aromatic rings in their backbone.

Structure within a Molecule

The important structural features within a polymer chain involve stereoisomerism, branching, molecular weight, end groups and impurities, and copolymerization. All these affect the arrangement between the polymer chains within a material, which affects its properties.

Stereoisomerism refers to the way that side groups are arranged in a polymer chain. It occurs in materials for which it is possible for mer units to be added to the growing polymer molecule in such a way that the side groups can be arranged in a random or orderly fashion. This is illustrated in Figure 4.27. The orientation of the side group with respect to the backbone can be (1) random (atactic), (2) on the same side (isotactic), and (3) on alternate sides (syndiotactic). Atactic polymers tend to be rubbery, whereas an isotactic polymer has more stiffness. The stereoisomerism within a polymer affects the ability of the chains to arrange themselves and pack together within the material, affects its stiffness. This is discussed in more detail later in this section.

Many thermoplastics are composed almost completely of linear chains. However, some thermoplastics have chains with branches. Branches can have short or long lengths and can occur rarely or frequently along a chain. The branches prevent close packing of the polymer chains. As a result, for a polymer of a given molecular weight, a more highly branched polymeric material has a lower density compared to a less branched or a linear polymer.

Branching interferes with the chains' ability to align and also interferes with bonding between chains. This enables chains to move more easily past each other compared to a linear molecule, resulting in a more flexible material compared to one with no branching. Reduced alignment results in a reduction in the glass transition temperature and reduced strength, rigidity, and dimensional stability, with other factors (e.g., molecular weight) being constant. Dimensional stability refers to the changes in the physical dimensions of a polymer material with changes in the temperature. Increased branching also decreases a polymer's ability to conduct heat; the heat must follow a longer, more winding path through the material because there are fewer intersections between the molecules. Finally, branching results in a decrease in the viscosity of a molten material, which is significant for component fabrication.

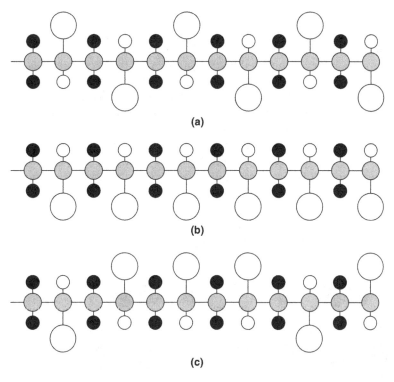

FIGURE 4.27

Schematic of stereoisomerism. (a) On alternate sides (syndiotactic). (b) On the same side (isotactic). (c) Random (atactic).

For a specific polymer, the average molecular weight and molecular weight distribution have an impact on physical and mechanical properties such as the tensile strength, impact toughness, creep resistance, and melting temperature. As mentioned earlier, the molecular weight is proportional to the length of a polymer chain. When the chains are short, the mechanical strength is lower compared to the same material with longer chains. This is because the chain ends act as impurities and disrupt alignment between chains. Greater alignment enables chemical bonding between chains, which reduces the flexibility and increases the strength of the material compared to a material with less chain alignment. Under an applied mechanical load, the chains in a shorter chain material slide more easily over each other compared to the same material with longer chains. Furthermore, longer chains result in more entanglement, which results in a greater resistance to the chains sliding over each other.

In addition to affecting the mechanical properties of a polymer, the average chain length and distribution of chain lengths affects a material's processing properties. Shorter chain materials flow better because there is less chain entanglement and chains can slide over each other more easily than in a longer chain material.

It is possible to have impurities polymerized into a polymer chain. Impurities have different chemical properties from the rest of the chain and can act as sites for decomposition, cross-linking, or other chemical reactions.

Copolymerization involves the combination of different types of mer units to form a polymer. If only one type of mer unit is used, the resulting molecule is called a *homopolymer*. Polymers with enhanced properties are obtained by combining different types of mers. In this case, the polymer is called a *copolymer*. The ways in which two different monomers can be combined are shown in Figure 4.28. The arrangement of the mer units can be random, alternating, block in a single chain, or as one chain grafted onto another. An example of a copolymer is high-impact polystyrene. This material is a graft copolymer that consists of polystyrene and polybutadiene rubber. Polystyrene is hard and inflexible, whereas polybutadiene is flexible. Adding the polybutadiene to the polystyrene results in a material with improved impact toughness compared to polystyrene. This copolymer material is used for toys, electronics, and appliances.

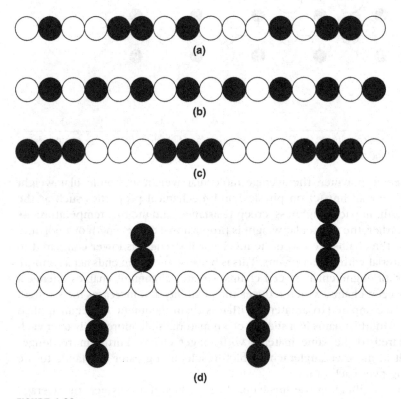

FIGURE 4.28

Different arrangements of a copolymer with two different mer units. (a) Random. (b) Alternating. (c) Block. (d) Graft.

Structure Between Molecules

The structure between polymer chains in a thermoplastic material results in materials referred to as amorphous or crystalline. In *amorphous* polymers the chains are entangled in a random arrangement, as shown in Figure 4.29(a). In *crystalline* polymers, the sections of a chain arrange in a regular pattern consisting of closely packed and folded chains, as shown in Figure 4.29(b). The folded chains are held together by localized chemical bonds. The crystalline regions are connected by amorphous regions. This definition of crystalline is different than that used for metals and ceramics. Examples of crystalline polymers include high-density polyethylene (HDPE), polypropylene (PP), and nylon.

The amount of crystallinity within a polymer varies. Some polymers (e.g., polycarbonate and polysulfone) are completely amorphous. Polymers with a structure that favors crystallinity are referred to as crystalline polymers, even though the actual crystallinity in the particular material is less than 100%. The formation of crystalline regions within a material occurs as a molten polymer cools below the melting point of the crystalline phase. Crystalline polymers are never completely crystalline because of the enormous number of chain entanglements in the melt, and it is impossible to achieve the amount of organization required to form a 100% crystalline polymer during cooling.

The degree of crystallinity has a significant impact on the properties of a polymer. Table 4.10 contains the properties of polyethylenes of varying degrees of crystallinity, and the data show that the melting point temperature, hardness, and tensile modulus increases as the degree of crystallinity increases. The melting point increases because more energy, in the form of heat, must be put into the polymer to break the chemical bonds that join the chains in the crystalline regions. The hardness and tensile modulus increase because of the increased stiffness of the polymer in the crystalline regions compared to the amorphous regions. Finally, as the crystallinity of a thermoplastic increases, the chemical resistance increases

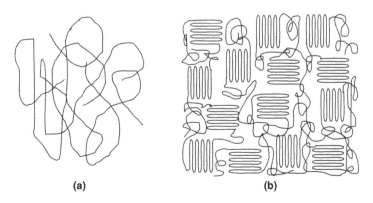

(a)   (b)

FIGURE 4.29

Structure between chains in polymers. (a) Amorphous. (b) Crystalline.

Table 4.10 Properties of Polyethylenes of Varying Degrees of Crystallinity

Property	Low Density	Medium Density	High Density
Density range (g/cm^3)	0.910–0.925	0.926–0.940	0.941–0.96
Crystallinity (approximate %)	42–53	54–63	64–80
Melting temperature (T_m) (°C)	110–120	120–130	130–136
Hardness (shore D)	41–46	50–60	60–70
Tensile modulus (MPa)	97–260	170–380	410–1240

because the denser packing of the chain molecules makes it difficult for a solvent or other chemical substance to penetrate.

Increases in the degree of crystallinity of a crystalline thermoplastic material will affect the material's toughness. In a polymer with a low crystallinity, at temperatures below the glass transition temperature an increase in crystallinity will result in increased stiffness and yield strength at the expense of toughness. For moderately crystalline materials at temperatures greater than the glass transition temperature, an increase in crystallinity will improve toughness. For a highly crystalline material at temperatures above and below the glass transition temperature, an increase in crystallinity will reduce toughness.

The degree of crystallinity is affected by processing variables such as the rate of cooling from the melt and the melt temperature. However, the maximum degree of crystallinity depends on the structure within the polymer chain. Anything that allows the polymer chains to form close-packed folded regions will improve the amount of crystallinity. Because chain mobility is required to form ordered structures, polymers with rigid backbones cannot crystallize under normal processing conditions. Polymers without large, bulky side groups have higher crystallinity than polymers with bulky side groups because the bulky side groups cannot get close enough together to form chemical bonds. In general, isotactic and syndiotactic polymers will crystallize to form a stiff material. Atactic polymers are rubbery, amorphous materials because the irregular side group arrangement of atactic polymers prevents the chains from forming crystalline regions. Finally, the molecular weight of a polymer can be made too high for some materials, where the long chains and increased degree of branching reduces the amount of crystallinity as the branches interfere with the formation of close-packed chain segments.

Amorphous thermoplastics exhibit a glass transition temperature above which the amorphous regions become mobile, and the elastic modulus decreases dramatically. Crystalline polymers exhibit both a glass transition temperature and a higher melting point temperature. At the glass transition temperature the amorphous regions start to soften, and at the melting point temperature the crystalline regions melt and become disordered random coils.

The structure between chains in thermosets consists of three-dimensional networks that form as a result of strong chemical bonds between chains and is referred to as cross-linking. The formation of the cross-linked structure involves a curing process, which can require the application of heat and possibly pressure. Cross-linking restricts the movement of atoms in thermosets, and the degree of cross-linking has an effect on a polymer's thermal properties. For a particular thermoset material, as the degree of cross-linking increases the temperature at which the material softens also increases. Furthermore, because of the strong chemical bonds between chains, thermosets remain strong until they break down chemically by charring or burning. Thermosets are seldom crystalline because cross-linking inhibits the ability of the chains to arrange themselves into close-packed regions.

In thermosets cross-linking improves dimensional stability at temperatures greater than the transition temperature of glass. However, high levels of cross-linking lead to embrittlement and a loss of toughness. Thus, there is a trade-off between obtaining a high glass transition temperature along with good toughness in a thermoset.

The magnitude of a polymer's glass transition temperature depends on the flexibility of the polymer chain and the bond strength between chains. The more flexible and the less bulky the mer unit, the easier it is for cooperative motion of mer units to occur, and thus the lower the glass transition temperature. However, strong bonds between chains will interfere with the motion. This is why thermosets have higher average glass transition temperatures compared to thermoplastics. Thermoplastics with the highest glass transition temperatures have stiff bulky chains and stronger bonds between chains. The stiff, bulky chains prevent motion in the amorphous regions as temperature increases.

Defects do arise in the structures along chains and between chains. These defects include low molecular weight components and impurities. During crystallization of a polymer that is cooling from the molten state, all the low molecular weight components and impure species are packed into the spaces between the crystalline regions. This reduces the strength and modulus of the amorphous regions.

More information about the chemistry, structure, and properties of polymers can be found in Billmeyer (1984), Rodriguez (2003), and Strong (2005).

4.7 COMPOSITES

A composite is a material system composed of a combination of two or more distinct materials for the purpose of obtaining specific features and properties. The distinct materials, which can be metallic, ceramic, or polymeric, retain their identity and can be physically identified from each other. Composite materials are developed to meet the requirements for an application where no single, homogeneous material is well suited. Thus, a composite material has properties that are superior to either constituent material alone.

The following are the four general categories of composites based on their physical construction.

1. Particles, flakes, spheres, or whiskers of one or more materials embedded in another material—called the *matrix*. Both the matrix and reinforcements are available as polymers, metals, and ceramics. Flake reinforcements are used where anisotropic mechanical behaviors are desired. As long as the flakes are parallel, uniform properties are obtained in the plane of reinforcement.

2. Continuous or discontinuous fibers embedded in another material. Both the fiber and matrix are available as polymers, metals, and ceramics. Discontinuous fibers are long or short, and their alignment can be unidirectional, bidirectional, or random.

3. Filled composite that is composed of a continuous porous or honeycomb matrix that is filled in with another material.

4. Sandwiches, laminates, and woven fibers of different compositions. Laminates are layered composites that consist of two or more different layers bonded together; layers can differ in material, form, or orientation. Laminates that consist of layers of composites result in increased bending strength compared to a composite material of similar total thickness that contains fibers in just one direction. Sandwiches are special-case laminates that consist of a thick low-density core between thin sheets of comparatively higher-density material.

Schematics of the different forms of composites are shown in Figure 4.30. Examples of composites and their applications are listed in Table 4.11. Figure 4.31 shows cross-sections of some different composites.

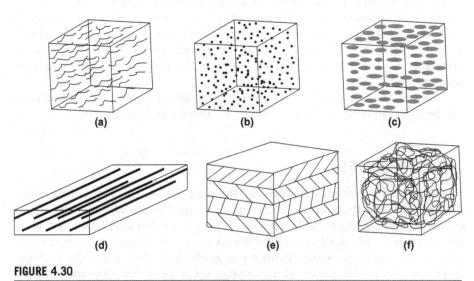

FIGURE 4.30

Schematics of the different forms of composites. (a) Discontinuous fiber. (b) Particulate. (c) Flake. (d) Continuous fiber. (e) Laminar. (f) Filled.

Table 4.11 Examples of Composite Materials and Their Applications

Composite Material	Applications
Plastic with short glass fibers	Housings for electronics; electrical connectors
Rubber with nylon and rayon fibers	Automobile tires
Epoxy with metal flakes	Electrically conductive adhesives
Plastic with long glass fibers laminated to copper	Electrical circuit boards
Paint with metal flakes	Conductive paints for electronics
Epoxy with boron nitride (BN) or aluminum nitride (AlN) particles	Thermally conductive dielectrics for electronic circuits

From a matrix material perspective, there are three categories of composites: organic matrix, metal matrix, and ceramic matrix.

Organic matrix composites include polymer-matrix composites and carbon-matrix composites. Both thermoset and thermoplastic resins are used for matrix materials and are filled with discontinuous glass or metal fibers, continuous glass fibers, and glass and metal flakes, whiskers, and particles.

Metal matrix composites include aluminum alloys, magnesium alloys, and titanium alloys as matrix materials and can be filled with ceramic fibers, whiskers, and particles.

Ceramic matrix composites contain particles, flakes, whiskers, and spheres.

In the case of composites developed for their mechanical properties, the improved mechanical properties generally result from a load-sharing mechanism between the matrix and reinforcement.

The metal, ceramic, and polymer materials that make up a composite material have been discussed in previous sections. In addition to the component materials used to make a composite, the interface between the matrix and embedded or reinforcement material is important to the properties of a composite. The interface must be properly controlled in order for a composite material to have the desired properties and resistance to environmental degradation. For example, in structural composites, the interface functions to transmit forces between the matrix and reinforcement. The desired bonding between the matrix and reinforcement may be weak, strong, or somewhere between depending on the application and the materials that make up the composite. The interfacial bonds between the matrix and embedded or reinforcement materials are the result of various mechanisms such as adsorption and wetting, electrostatic attraction, chemical bonding, and reaction bonding (Kim & Yiu-Wing Mai, 1998). There can be more than one

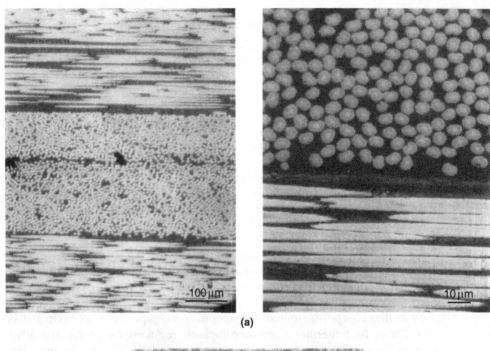

(a)

(b)

FIGURE 4.31

Cross-sections of different composites. (a) Laminate of carbon fiber in epoxy (Chawla, 1998). (b) SiC whiskers in alumina (Chawla, 1998).

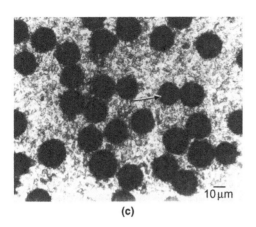
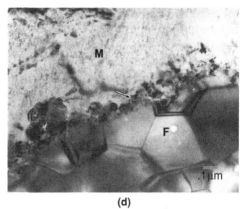

(c)   (d)

FIGURE 4.31   *Continued*

(c) Alumina fibers in a magnesium alloy. (d) Same as (c) but at a higher magnification showing the reaction between the metal matrix and the alumina fiber. (*Source:* Parts (a) and (b) are reprinted with the kind permission of Springer Science and Business Media.)

bonding mechanism for a particular combination of matrix and reinforcement. A larger discussion of interfaces occurs later in this chapter.

The surface properties of embedded materials affect the bond to the matrix and the properties of the interface region. The surface area of the embedded materials can be much greater than its geometrical value because of pores, pits, or cracks present on the surface. This increased surface area improves the mechanical bonding between the matrix and the embedded material. Surface treatments can add surface chemical groups or remove the original surface, giving rise to a chemically and structurally different region. Exposure to air before composite processing can result in the adsorption of chemical species, which may alter or eliminate certain beneficial surface reactivity. Adsorbed materials may also desorb at the elevated temperatures seen in composite fabrication and can be a source of volatiles, which, if not removed, can be the origin for voids that disrupt the interface.

Modifications to the surface of an embedded material influence reactions with its matrix material. This is important because acceptable interfacial interaction between the reinforcement and the matrix is determined by the reactivity between the reinforcement surface and that of the matrix. Sufficient knowledge of the interface region and its effect on reinforcement-matrix adhesion and composite mechanical performance has been achieved so that the reinforcement-matrix interface region can be engineered through the use of reinforcement surface treatments and coupling agents in order to optimize composite performance.

The chemical and microstructure characteristics of the interface region affect the performance of the composite in terms of its mechanical strength and chemical and thermal durability. Defects, such as unreacted matrix components and impurities, can diffuse to the interface region and alter the local structure. This interferes with intimate contact between reinforcement and matrix, producing a material that does not have useful properties. Furthermore, the interface region formed during composite processing may not be in its equilibrium configuration as a result of processing constraints. This will affect a composite's behavior.

Glass reinforcement materials are sometimes treated with coupling agents to promote greater adhesion with specific polymer resins. The ideal coupling agent should provide a flexible layer at the interface that will improve adhesive strength of the fiber-resin bond and reduce the number of voids at the interface. Voids around fibers are harmful because under load the unsupported section of fiber may buckle and transfer stress to the polymer matrix.

In structural ceramic matrix composites, the interface between the matrix and whisker reinforcements is intentionally weakened, resulting in reduced mechanical coupling between the matrix and reinforcement. When this type of composite is deformed, cracks propagate through the matrix until they meet a whisker. The weak interface allows the whisker to pull out from either side of the advancing crack instead of continuing through the whisker. The friction from this pullout dissipates a large amount of energy. This behavior results in substantial toughening of the ceramic. The bond strength must be controlled so that it is not so large that whisker pullout is prevented or so low that whisker pullout does not produce much energy absorption.

4.8 SURFACES

The surface of a solid material can be viewed as a distinct material with properties that depend on its composition and structure. The surface of a solid material consists of the top few atomic layers and has different chemical and electrical properties compared to the bulk material. Many of the electronic properties of a bulk material depend on the three-dimensional periodic structure inside the solid. The loss of periodicity in one dimension at the surface results in surface electronic properties that are different from the bulk. The lack of neighboring atoms on one side of the surface atoms may result in available chemical bonds, which "dangle" into the space outside the solid and are available for chemical reaction. This is illustrated in Figure 4.32.

The surface composition of a material is usually different than the composition within the bulk of the material. This is usually a result of the oxidation of the surface when it is exposed to air. Most metals readily form a thin surface oxide in air at room temperature. Exposure to hydrocarbons and other airborne pollutants results in the deposit of thin layers of the materials on surfaces. Also, substances, such as oils or dirt, can be present on the surface of a material as a result of handling or improper cleaning after processing.

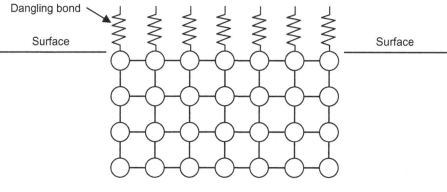

FIGURE 4.32

Diagram of dangling bonds at a surface.

Surface composition and cleanliness affects the bond strength of adhesives and coatings; the electrical contact resistance between two metal conductors; and the ability to form good weld, solder, and braze joints to a material. It also influences the effects of a corrosive environment on a material and the manner in which two materials wear when rubbed against each other.

The structure of a surface is related to its macroscopic and microscopic roughness. Even surfaces that appear smooth to the naked eye have bumps and undulations on a microscopic level. The peak height of surface roughness can be greater than 1 mm or as small as the spacing between atoms, about 0.5 nm. Surfaces can also have pits formed by chemical or mechanical means.

Surface structure affects the bond strength of adhesives and coatings that adhere by mechanically locking themselves to the surface. Surface roughness also affects the electrical contact resistance between two conductors since the points of contact occur only at the surface peaks, not along the entire contact surface.

4.9 INTERFACES

An interface is the boundary where two materials meet each other. Interfaces occur in a variety of material systems such as in composites between the matrix and embedded materials, in joints between the joint filler and each of the components being joined, and between a material and a surface coating.

There are several types of interfaces including abrupt, diffusion, and compound. Each of these is shown in Figure 4.33. The type of interface that forms depends on the materials in contact, the surface composition of the materials before contact, and the processes used to bring the materials in contact. The type of interface that forms between two materials affects the mechanical, electrical, and thermal properties of the interface and the reliability of the interface.

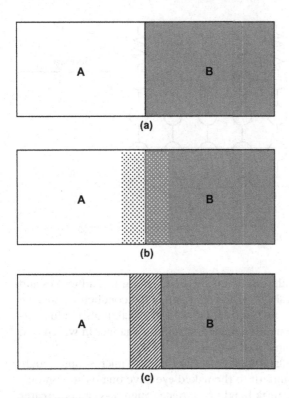

FIGURE 4.33

Schematics of interfaces. (a) Abrupt. (b) Diffusion. (c) Compound.

4.9.1 Abrupt Interfaces

An abrupt interface is characterized by an abrupt change from one material to the other in a distance of the order of the atomic spacing (i.e., 0.2–0.5 nm). A diagram of this type of interface is shown in Figure 4.33(a). Abrupt interfaces occur when there is neither diffusion nor reaction between the materials at the interface. Examples of abrupt interfaces are those between matrix and filler in polymer matrix composites, between an adhesive and adherend in some adhesive joints, and between some coatings and base materials.

Even though there are no reactions, there are chemical and electrical interactions between the two materials at an abrupt interface. These interactions involve the atomic bonding forces that exist between the mating materials and which hold them together.

A variation of the abrupt interface is the mechanical interlocking abrupt interface. The mechanical interface is an abrupt interface on a rough surface where an adhesive or coating conforms to the roughness and the rough surface is "filled in" to give mechanical interlocking.

4.9.2 **Diffusion Interfaces**

The diffusion interface is characterized by a gradual change in composition across the interfacial region with no compound formation. A diffusion interface forms when two materials are brought together and heat and pressure are applied to the interface, along with mutual solid solubility between the two materials. The temperature, pressure, and time must be sufficient to allow diffusion of the atoms from the materials into each other. A diagram of a diffusion interface is shown in Figure 4.33(b). The extent of diffusion depends on time and temperature. This type of interface is found in metallic systems.

The mixing that occurs as a result of the diffusion results in increased adhesion between the two materials. However, if the diffusion of atoms is much greater in one direction than the other, then there will be void formation in the material losing the excessive atoms. The voids weaken the material.

4.9.3 **Compound Interfaces**

A compound interface contains one or more layers of compounds between the two materials that have been brought together. The compounds form either through chemical, metallurgical, or solid-state reactions. A schematic of a compound interface is shown in Figure 4.33(c). The compounds formed are often brittle. The compound interface is generally a contributor to good adhesion between the materials that have been brought together. But if the reaction region is too thick, the development of porosity and the formation of microcracks can lead to poor adhesion.

4.10 **DEFECTS**

A wide variety of defects can form in a product element's structure. The specific defects that form, their size, and their number depend on the particular materials and manufacturing process used to form a product element. The impact of defects is to reduce the reliability of a product element. However, this is not to say that no defects can be present. Instead, it is important to understand the defects that are tolerable for a particular product element.

Examples of the defects that can form are discussed in Chapter 5 for some specific manufacturing processes.

4.11 **MATERIALS INFORMATION RESOURCES**

There are many valuable sources of information for materials properties, features, and applications. The different types of sources are listed next.

4.11.1 Handbooks and Textbooks

Thousands of handbooks and textbooks discuss all types of materials, their properties and microscopic structures, applications, reliability, and fabrication methods. Such books are available at public, college, and university libraries. They can also be purchased at bookstores and online. Some excellent handbooks are available through technical societies such as ASM International, the Society of Plastics Engineers, the American Ceramics Society, NACE International, and the Society of Manufacturing Engineers, just to name a few. A $150 book is well worth the price if it provides information that leads to better and faster decisions that save time and money. Examples of these handbooks and textbooks are referenced throughout this book.

4.11.2 Journals

Every month hundreds of science and engineering journals are published that contain articles about experiments. The articles are peer reviewed, which means that each article is anonymously reviewed by the authors' peers in the specific field of science or engineering to which the article relates. The purpose of such a review is to ensure that the content and scientific methods used meet acceptable standards.

Many of the journals go back several decades, and a few go back to the late 19th century. Recent editions of a number of journals are available both in print and electronically through the Internet. Universities also have access to searchable databases that contain reference information about specific papers dating back at least to 1970.

Of all the published journals, more than 50 are devoted to materials science and engineering. Others focus on other areas of engineering (e.g., mechanical, electrical, chemical), where materials are discussed with respect to characterizing a particular property. Lists of examples of both types of journals are shown in Table 4.12. Many of the publishers of these journals provide Internet search engines for locating online articles of interest.

4.11.3 Standards

The purpose of standards is to standardize engineering practices, and they are developed and written by committees of engineers and scientists within the industry or industries to which the standards apply. Standards address the following: (1) test methods for evaluating the properties of materials, (2) required properties of standard materials, (3) materials that should be used in specific applications, (4) required properties of materials used in specific application, (5) engineering design requirements, (6) product performance and safety requirements, and (7) test methods for evaluating the performance and reliability of products.

Various organizations write and publish standards. Table 4.13 shows a list of some of these organizations and their engineering focuses.

Table 4.12 Selected Journals with Materials Science and Engineering Information

Materials-Focused Journals	Journals with Other Engineering Focuses
Acta Materialia	*Catalysis Reviews: Science and Engineering*
Advanced Materials	*IEEE Transactions on Advanced Packaging*
Advanced Materials & Processes	*IEEE Transactions on Device and Materials Reliability*
Advances in Cement Research	*IEEE Transactions on Dielectrics and Electrical*
Applied Surface Science	*Insulation*
Biopolymers	*IEEE Transactions on Semiconductor Manufacturing*
British Ceramic Transactions	*Journal of Applied Physics*
Cement & Concrete Composites	*Journal of Biomedical Materials Research*
Journal of the American Ceramic Society	*Journal of Physics*
Ceramic Engineering and Science	*Journal of the Electrochemical Society*
Corrosion	*Journal of Tribology*
Corrosion Science	*Mechanical Engineering*
Glass Technology	*Philosophical Magazine*
Journal of Applied Polymer Science	*Physical Review*
Journal of Electronic Materials	*Sensors and Actuators*
Journal of Engineering Materials and	*Tribology & Lubrication Technology*
Technology	*Tribology Transactions*
Journal of Materials Science	*Wear*
Materials Science and Engineering	
Metallurgical and Materials Transactions	
Plastics Engineering	
Polymer Engineering and Science	
Polymer International	
Scripta Materialia	
Surface Science	
Thin Solid Films	

4.11.4 Technical Societies

A number of technical societies focus on materials. There are also technical societies that address materials and materials engineering, but from the perspective of other engineering disciplines. All these societies publish textbooks, handbooks, standards, and journals; offer classes; and have technical meetings. Some of these societies and their focuses are shown in Table 4.14.

4.11.5 Patents

The U.S. Patent and Trademark Office (USPTO) provides an online searchable database (*www.uspto.gov*) of all United States patents going back to 1799. Patents for materials describe the engineering rationale that justifies the creation of the material. Patents for mechanical and electrical devices sometimes discuss the materials used in the devices. Patents also help provide an understanding of when and how specific materials were commercialized.

Table 4.13 Selected Standards Organizations and Their Technical Focus

Organization	Focus
ASTM International	Materials, product, processing, and characterization standards for a wide range of materials, products, and industries
National Electrical Manufacturers Association	Electrical equipment design and materials
American Gear Manufacturers Association	Gear materials, design, and fabrication
International Organization for Standardization (ISO)	A wide variety of industries, applications, and materials
ASME (mechanical engineering)	Materials and components for pressure vessels
American Welding Society (AWS)	Welding materials, procedures, and weld characterization.
SAE International	Components, materials, and design for ground vehicle and aerospace applications

Table 4.14 Technical Societies and Their Focus

Society	Focus
ASM International	Metals, coatings, fabrication, materials characterization
Society for Plastic Engineers (SPE)	Plastics
ACerS	Ceramics
American Welding Society (AWS)	Welding
NACE	Corrosion engineering
Society for Automotive Engineers (SAE)	Automotive, heavy duty vehicles, and aerospace
Institute for Electrical and Electronic Engineers	Electrical equipment, electronics, materials for electronics
International Microelectronics and Packaging Society (IMAPS)	Materials and manufacturing methods for electronics
Surface Mount Technology Association (SMTA)	Materials and manufacturing methods for electronics
AVS	Materials and processing related to vacuum science

4.11.6 **Technical Data from Manufacturers**

Suppliers of materials, components, and subassemblies usually provide technical information about their products in the form of technical data sheets. For materials (e.g., metal bar stock, adhesives, and paint), the information is related to their composition, their properties, and typical applications where they can be used. It must be kept in mind that the materials' properties data reported are usually just nominal values. Variation of properties should be expected from batch to batch of material.

Component and subassembly manufacturers provide information about components' electrical, mechanical, optical, cosmetic, and physical properties. Sometimes, they also provide information about the use conditions for which a component or subassembly is designed. Data sheets may or may not provide information about the materials used in a component. As with the materials information, the data about components are usually nominal values.

4.11.7 **The Internet**

It is important to be skeptical of this general resource. People have become accustomed to searching for information on the Internet and accepting what is found without reservation. As a result, many engineering teams rely on it for information that will guide design and manufacturing decisions. The quality of the source should be considered; remember, anyone can post information on the Internet. The other resources just listed here are much more reliable.

REFERENCES

Aramaki, S., and R. Roy, *Journal of the American Ceramic Society*, 45:229, 1962.

ASM Handbook, Volume 2: Properties and Selection: Nonferrous Alloys and Special-Purpose Materials, p. 133. ASM International, 1990.

Avedesian, M., and H. Baker (eds.), *Magnesium and Magnesium Alloys*, ASM International, 1999.

Barrett, C., and T.B. Masalski, *Structure of Metals*, Third Edition, Pergamon Press, 1980.

Barsoum, Michel, *Fundamentals of Ceramics*, McGraw-Hill, 1997.

Billmeyer, F.W., Jr., *Textbook of Polymer Science*, Third Edition, John Wiley & Sons, 1984.

Brandes, E.A., and G.B. Brook (eds.), *Smithells Metals Reference Book*, Seventh Edition, Butterworth-Heinemann, 1992.

Chawla, K.K., *Composite Materials: Science and Engineering*, Second Edition, Springer-Verlag, 1998.

Chen, C.W., *Magnetism and Metallurgy of Soft Magnetic Materials*, Dover Publications, 1986.

Coble, R.L., and J.E. Burke, *Progress in Ceramic Science*, J.E. Burke (ed.), 3:197 (Figure 25), Pergamon Press, 1963.

Davis, J.R. (ed.), *Stainless Steels*, ASM International, 1994.

Davis, J.R. (ed.), *Aluminum and Aluminum Alloys*, ASM International, 1993.

Davis, J.R. (ed.), *Copper and Copper Alloys*, ASM International, 2001.

Davis, J.R. (ed.), *Nickel, Cobalt, and Their Alloys*, ASM International, 2000.

Donachie, M.J., and S.J. Donachie, *Superalloys: A Technical Guide*, Second Edition, ASM International, 2002.

Donachie, M.J., *Titanium: A Technical Guide*, Second Edition, ASM International, 2000.

Doremus, R.H., *Glass Science*, Second Edition, Wiley, 1994.

Engineered Materials Handbook, Volume 4: *Ceramics and Glasses*, ASM International, 1991.

Guy, A.G., *The Essentials of Materials Science*, McGraw-Hill, 1976.

Haasen, P., *Physical Metallurgy*, Third Edition, Cambridge University Press, 1996.

Humpston, G., and D.M. Jacobson, *Principles of Soldering*, ASM International, 2004.

Hwang, J.S., *Modern Solder Technology for Competitive Electronics Manufacturing*, pp. 196-197, McGraw-Hill, 1996.

Kim, J.-K., and Y.-W. Mai, *Engineering Interfaces in Fiber Reinforced Composites*, Elsevier, 1998.

Kingery, W.D., H.K. Bowen, and D.R. Uhlmann, *Introduction to Ceramics*, Second Edition, John Wiley and Sons, 1976.

Lampman, S. (ed.), *Characterization and Failure Analysis of Plastics*, ASM International, 2003.

Lee, W.E., and W.M. Rainforth, *Ceramic Microstructures: Property Control by Processing*, Chapman and Hall, 1994.

Llewellyn, D.T., *Steels: Metallurgy and Applications*, Butterworth-Heinemann, 1994.

McCurrie, R.A., *Ferromagnetic Materials: Structure and Properties*, Academic, 1994.

Metals Handbook, Volume 1, Properties and Selection: Irons, Steels, and High-Performance Alloys, Tenth Edition, ASM International, 1990.

Metals Handbook, Volume 2, Properties and Selection: Nonferrous Alloys and Special Purpose Materials, Tenth Edition, ASM International, 1990.

Porter, D.A., and K.E. Easterling, *Phase Transformations in Metals and Alloys*, Van Nostrand Reinhold, 1989.

Reed-Hill, R.E., and R. Abbaschian, *Physical Metallurgy Principles*, Third Edition, PWS-Kent Publishing Co., 1992.

Richerson, D.W., *Modern Ceramic Engineering: Properties, Processing and Use in Design*, Third Edition, CRC Press, 2006.

Rodriguez, F., *Principles of Polymer Systems*, Fifth Edition, Taylor & Francis, 2003.

Shelby, J.E., *Introduction to Glass Science and Technology*, Second Edition, Royal Society of Chemistry, 2005.

Strong, A.B., *Plastics: Materials and Processing*, Third Edition, Prentice Hall, 2005.

Van Vlack, L., *Physical Ceramics for Engineers*, Addison-Wesley, 1980.

Verhoeven, J.D., *Fundamentals of Physical Metallurgy*, Wiley, 1975.

Manufacturing Process Considerations

5.1 INTRODUCTION

A wide variety of manufacturing processes are used to form product elements. These processes use thermal, mechanical, electromagnetic, chemical, electro-chemical, and radiation means to manipulate materials into product elements with the desired shape, dimensions, material features, and material properties. The manufacturing methods used to manipulate materials into a product element have a direct impact on the composition, microscopic structure, and defects within the materials that make up the product element. The properties and reliability of the materials in a product element depend only on these three items.

Obtaining the desired composition, microscopic structure, and set of defects in a product element requires control over the manufacturing processes. Therefore, it is important for a design team to understand the effects of manufacturing processes on a material's composition, microscopic structure, and defects.

This chapter first reviews manufacturing processes and then goes on to discuss the various aspects of manufacturing that must be considered in order to consistently form product elements that satisfy their performance and reliability requirements.

5.2 COMPONENT FABRICATION PROCESSES

For the purposes of our discussion, two broad categories of component fabrication processes are defined: primary and secondary. Many components are fabricated using a combination of these processes. The term *component-in-process* refers to an incomplete component that awaits further processing. *Primary processes* are used to shape a material into a form that is at or close to the final shape and dimensions of the component being fabricated. *Secondary processes* are used to modify the surface of a component-in-process, to modify the material properties of a component-in-process, and to deposit a coating on a component-in-process.

Secondary processes are used when no combination of materials and primary processes enables the fabrication of a component that meets all of its design requirements. When secondary processes are used, the component-in-process after primary processing is referred to as the *base material*. For example, aluminum oxide is the base material in a component composed of sintered aluminum oxide with a metal coating. A steel screw with zinc coating has steel for the base material.

The rest of this section discusses some of the various manufacturing processes available. However, many processes have been omitted. References for more detailed information are provided.

5.2.1 Primary Processes

Most primary processes can be placed into one of the following categories:

1. Molding and casting
2. Powder compaction and sintering
3. Mechanical and thermomechanical deformation
4. Material removal
5. Lamination

A brief description of some of the processes in each of these categories follows. More information about these processes can be found in *ASM Handbook,* Volume 7 (1998); *ASM Handbook,* Volume 14a (2005); *ASM Handbook,* Volume 14b (2006); *ASM Handbook,* Volume 15 (1988); *ASM Handbook,* Volume 16 (1989); Berins (1991); Cubberly and Bakerjian (1989); DeGarmo, Black, and Kohser (2002); Kalpakjian and Schmid (2006); Reed (1995); Richerson (2006); Rosato and Rosato (1995); and Swift and Booker (2003).

Molding and Casting

These methods involve pouring or injecting molten material into a mold that is formed in the shape of the component to be fabricated. After the molten material has solidified and sufficiently cooled, it is removed from the mold. These processes can be used to form components from metals, polymers, and glass. For plastics, this category of processes includes injection molding and blow molding. For metals, it includes die casting, sand casting, and investment casting. Figure 5.1 shows schematics of some of these processes. The details of two processes will be discussed here as examples.

For plastic injection molding, thermoplastic pellets are supplied to the injection molding machine through a feed hopper. The screw in the barrel turns in order to move the material to the screw tip. The plastic is heated and melted as a result of (1) heat from the friction of the plastic moving against the screw and the sides of the barrel and (2) passing through heated zones in the barrel. The molten plastic is pushed forward. When there is enough molten plastic in front of the screw tip, the screw is pushed forward and the molten plastic is pumped

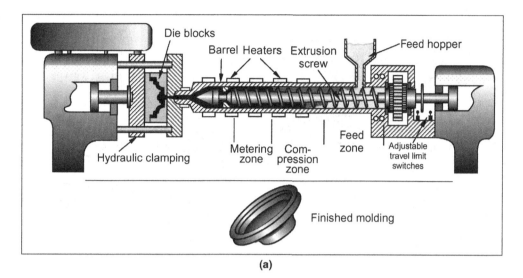

(a)

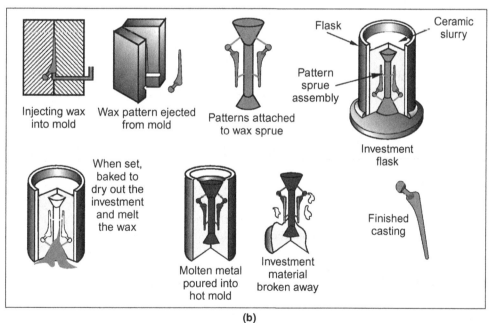

(b)

FIGURE 5.1

Schematics of various molding and casting processes. (a) Injection molding. (b) Investment casting. (*Source:* From Swift and Booker, 2003. Reprinted with permission of Elsevier.) (*Continued*)

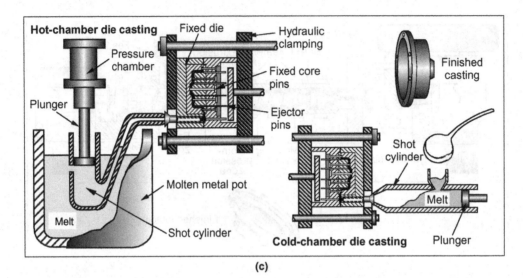

Hot-chamber die casting Fixed die Hydraulic clamping

Pressure chamber

Plunger

Fixed core pins

Ejector pins

Finished casting

Shot cylinder

Melt

Melt

Molten metal pot

Shot cylinder

Cold-chamber die casting Plunger

(c)

FIGURE 5.1   *(Continued)*

(c) Die casting.

into the mold cavity. Once the cavity is filled, the plastic cools. When the component is sufficiently solidified and cooled, the mold opens and the component is ejected.

For die casting, molten metal is forced into the die cavity (mold). After the component solidifies, the die opens and the component is ejected.

For any molding or casting process, it is necessary to control the composition of the materials used and the temperature of the molten material. Some of the processes also require control of other variables, such as the pressure used to force the material into the mold, the mold temperature, the time before opening the mold, and the application of a lubricant to the mold surface so that the component can be easily removed from the die/mold.

Powder Compaction and Sintering

These processes involve placing a metal or ceramic powder into a die and applying pressure to consolidate the powder into the desired shape. The compaction is followed by baking in a furnace at high temperatures during which the powder particles coalesce (i.e., sinter) to form a solid component. A schematic of the powder compaction process is shown in Figure 5.2. Hot pressing and hot isostatic pressing are processes that simultaneously compact and sinter the components.

Ideally, a sintered component has a controlled amount of porosity that provides the finished component with the required engineering properties. The amount of porosity and the size and spatial distribution of pores throughout a sintered component are functions of the compaction process and the sintering conditions.

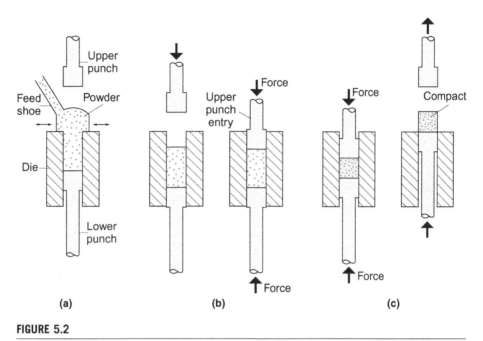

FIGURE 5.2

Schematic of the powder compaction process. (*Source:* Reprinted with permission of the Metal Powder Industries Federation.)

Mechanical Deformation

These processes involve mechanical deformation of a solid piece of metal or polymer to form it into the desired shape. The starting material can be in the shape of a sheet, tube, solid rod or wire, plate, or slab. In some processes, the material is heated so that it can be shaped more easily. Metal components can be stamped from a sheet of metal, drawn from a sheet of metal, or forged or extruded from a bar or slab of metal. Polymer components are formed by compression molding, and thermoforming. Figure 5.3 shows schematics of plastic compression molding, metal forging, and metal bending processes.

For compression molding, a plastic component is formed from a powder, pellets, or preform. A *preform* is a large pellet or plug of material that has been partially cured. As the mold closes, there is a shearing action of the material being compressed. Also, the two halves of the mold are heated. Both of these cause the plastic to soften and fill the mold cavity. Continued heating and pressure are applied to completely cure the plastic after forming.

Material Removal

These processes involve removing material from a plate, block, or rod of material. They are also used for additional processing of components that have been cast,

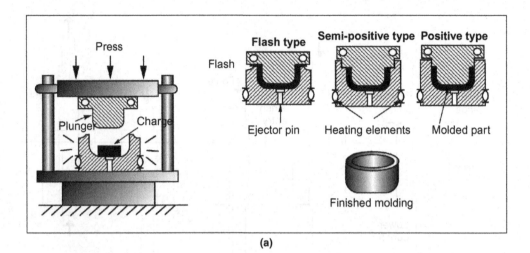

(a)

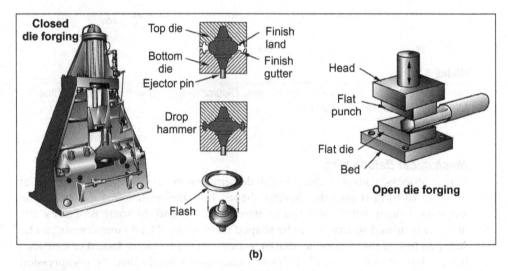

(b)

FIGURE 5.3

Schematics of mechanical deformation processes. (a) Plastic compression molding.
(b) Metal forging. (*Source:* Swift and Booker, 2003. Reprinted with permission of Elsevier.)

molded, or formed and require removal of excess material, finishing to tolerances that are tighter than would be possible through forming processes, or cosmetic alteration. Material removal processes include mechanical machining, chemical etching, and laser beam machining. Metals, polymers, and ceramics can be machined using some or all of these techniques.

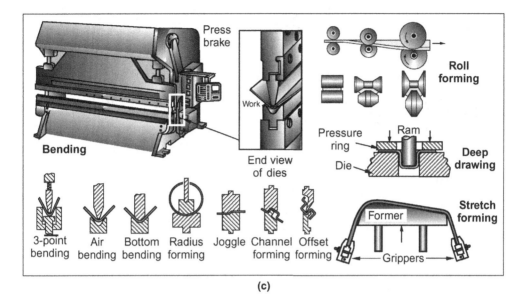

(c)

FIGURE 5.3 *(Continued)*

(c) Sheet metal forming.

Mechanical machining includes grinding, milling, and drilling. Chemical etching forms fine features in thin pieces of metal and etches away unwanted material from larger pieces of metal. Laser beam machining is used to drill holes in metals, polymers, and ceramics, and to cut these materials. Figure 5.4 shows schematics of drilling, milling, grinding, and lathing processes.

Lamination

This method involves assembling individual sheets of materials into a multilayered material or component. It is typically used to make composite materials and involves pressing sheets of materials together with or without the use of an adhesive between the layers. Sometimes heat is also applied. A lamination process is illustrated in Figure 5.5.

5.2.2 **Secondary Manufacturing Processes**

Secondary processes are used to modify a component-in-process and are able to be placed in one of the following categories:

- Modification of the material through the entire cross section of a component

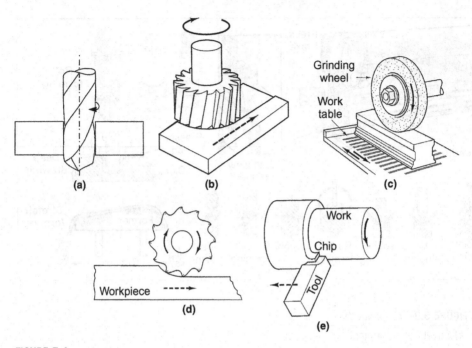

FIGURE 5.4

Schematics of material removal processes. (a) Drilling. (b) Milling. (c) Grinding. (d) Milling. (e) Lathing.

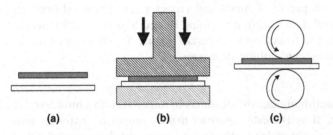

FIGURE 5.5

Schematic of a lamination process. (a) Materials to be laminated. (b) Press process. (c) Roller process.

- Modification of the surface of a component
- Deposition or growth of a coating onto the surface of a component

These processes are described in more detail next. More information about secondary processes can be found in *ASM Handbook,* Volume 4 (1991); *ASM Handbook,* Volume 5 (1990); Cubberly and Bakerjian (1989); DeGarmo et al. (2002); and Kalpakjian and Schmid (2006).

Modification of the Material through the Entire Cross Section of a Component

Heat treating is used to bring about desired changes in a metal's microstructure, with the associated changes in its properties. Different heat-treatment processes are used for purposes such as increasing the strength of metals, increasing the ductility of metals, or improving the magnetic properties of ferromagnetic alloys. Heat treatment processes involve controlled heating and cooling of components. The specific temperatures and treatment processes depend on the material and the microstructure and properties desired.

Steel alloys are strengthened and hardened by heating the material in an oven or furnace. After exposure to the high temperature, the steel can be cooled at different rates. For slow cooling, the furnace is turned off and the item is left inside or the item is removed from the furnace and allowed to cool in the air. Conversely, an item can be immersed in a liquid such as oil or water to obtain a fast cooling rate. For a given composition of steel, the microstructure depends on the heating temperature and the cooling rate. The heat treatment used depends on the steel composition, the shape and size of the component, and the desired microstructure and material properties.

Figure 5.6 shows the microstructures for a steel alloy that contained about 0.6 weight percent carbon. Figure 5.6(a) shows the microstructure before the heat treatment. The alloy consists of spheroidized iron carbide in a matrix of ferrite. This microstructure is desired in a steel plate or bar that must be easily formed into the desired shape using deformation processes. Figure 5.6(b) shows the same alloy after the component-in-process was heat treated. The microstructure consists of martensite. This strong, hard phase results in a material with high strength and hardness. The sample shown in Figure 5.6(b) is harder and stronger than the

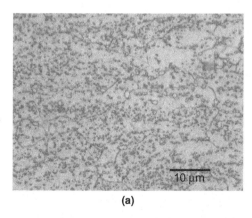

(a)

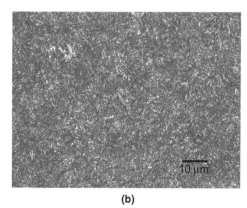

(b)

FIGURE 5.6

Micrographs of a steel alloy (a) before and (b) after heat treating. (Courtesy of Aston Metallurgical Services.)

sample shown in Figure 5.6(a). However, the sample in Figure 5.6(a) has more ductility, which enables the deformation required during forming.

Aluminum alloys, copper alloys, and nickel alloys are strengthened and hardened using a two-step process. First, the component is heated to an elevated temperature, held there for some period of time, and then quickly cooled. The purpose is to form a supersaturated solid solution in the alloy. This step is referred to as solution treatment. Then the component is heated, but to a lower temperature than used for the solution treatment. The component is held at this temperature for a period of time, during which second-phase particles form in the alloy. The component is then cooled to room temperature. The second step is referred to as precipitation hardening or aging. The strength and hardness of an alloy depends on the temperature and time for the second step.

Figure 5.7 shows micrographs of a permanent mold cast aluminum alloy with 6 weight percent silicon and 3.5 weight percent copper. Figures 5.7(a) and (b) show the as-cast and post-heat treated microstructures, respectively. The nominal yield strengths for the as-cast and heat treated samples are 130 MPa and 185 MPa, respectively.

Ferromagnetic alloys are heat treated at high temperatures to allow for grain growth and to allow gaseous impurities like hydrogen to diffuse out of the material. The resulting material has improved magnetic properties compared to the pretreated material.

Heat treatments like annealing and tempering increase the ductility of a metal while decreasing its strength and hardness. These processes involve heating a

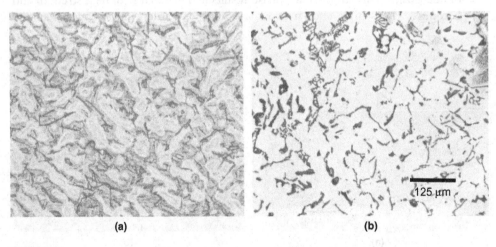

(a)   (b)

FIGURE 5.7

Micrographs of permanent mold cast aluminum alloy (a) as cast and (b) after heat treatment. (*Source: ASM Handbook,* Volume 9, 2004. Reprinted with permission of ASM International.)

metal to a moderately elevated temperature, which depends on the particular metal being heat treated.

Modification of the Surface of a Component

Chemical, mechanical, thermal, and thermochemical methods are used to modify the composition or microscopic structure of a component in a thin region near its surface, change the physical texture of the surface, or alter the chemical properties of the surface. These processes are used to increase wear resistance, enhance fatigue resistance, reduce the coefficient of friction, and improve the ability to form good weld, solder, braze, and adhesive joints to a component. The different types of surface modification processes are described next.

Surface heat treatment. These processes are used to form a strong, hard surface layer, referred to as the case, around a more ductile core of material. This is referred to as case hardening. The core is the inner portion of material not affected by the heat treatment. Processes such as induction hardening, laser hardening, and flame hardening are used to heat the near surface region of a steel component to the desired temperature. The heat treatment temperature and cooling rate are selected and controlled to obtain the desired microstructure and mechanical properties. The composition of the case and core remain the same, but the microstructures are different. For induction hardening, a high-frequency electric current is passed through a conducting coil surrounding the component to be heated. The electromagnetic energy created induces a current to flow near the surface of the component. After a certain period of time at the elevated temperature, the component is cooled. Figure 5.8 shows schematics of the flame and induction heat treating processes. Figure 5.9

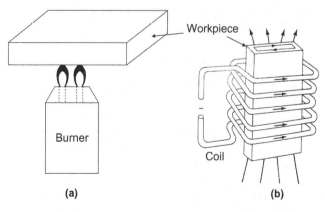

(a)   **(b)**

FIGURE 5.8

Schematics of surface heat-treating processes. (a) Flame hardening. (b) Induction hardening.

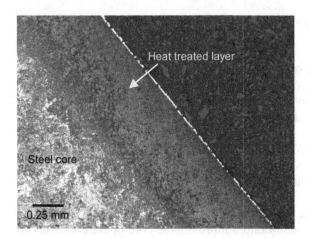

FIGURE 5.9

Micrograph of a steel alloy after laser surface heat treating. (Courtesy of Aston Metallurgical Service.)

shows the microstructure of a steel alloy after laser surface heat treating. The image shows a hard martensite phase at the surface and a softer material at the core of the component.

Thermochemical processes. These processes change the composition and microstructure in a thin region at the surface of a component. They involve heating a metal component to a high temperature in the presence of gases, liquids, or solids that contain elements such as carbon or nitrogen. At the high temperatures, these elements diffuse into or react with the material near the surface of the component being processed, creating a hard layer at the surface and improving the mechanical properties at the surface. Examples of thermochemical processes include steel carburizing, nitriding, and carbonitriding. An example of a carburized steel is shown in Figure 5.10.

Mechanical processes. Shot peening, grit blasting, and sanding are examples of processes used to mechanically modify the surface of a component. Shot peening modifies the residual stresses at the surface of a component to make it more resistant to fatigue. It involves spraying beads of metal, ceramic, or glass against a component's surface. Grit blasting cleans and roughens surfaces. It involves spraying a component's surface with fine particles made of a variety of materials including ceramics. Sanding removes excess material and dirt. It is also used to give the sanded surface a certain appearance. An example of a sanded metal surface is shown in Figure 5.11.

Chemical cleaning processes. These processes remove dirt, oils, and other contaminants from the surface of a component. Cleaning is done to prepare

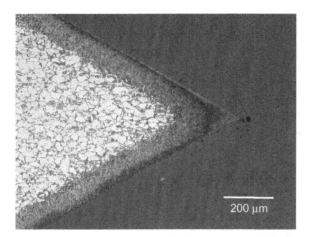

FIGURE 5.10

Micrograph of a carburized steel surface. (Courtesy of Aston Metallurgical Service.)

FIGURE 5.11

Metal components before and after surface sanding. (Courtesy of Timesavers, Inc.)

surfaces for subsequent processing or for cosmetic reasons. Acids, bases, detergents, and solvents are used for cleaning.

Coating a Component Surface

Surface coatings consist of thin layers of one or more materials that have been deposited onto or grown from the surface of a base material. The base material is called the *substrate*. Coatings range in thickness from less than 100 nm to hundreds of microns. A coating is grown from a surface by chemical reaction of the substrate surface with a vapor or liquid. Coatings or surface coatings provide wear, corrosion, chemical, electrical, optical, and cosmetic properties that exceed those possible for the base material surface.

The combination of a base material and surface coating enables the fabrication of components that meet performance, reliability, and cost requirements that would not be possible using an uncoated base material. For example, carbon steel screws are coated with zinc. The steel provides the strength, and the zinc provides corrosion protection. Without the zinc coating, the steel would corrode too quickly in products used outdoors.

Different coating processes are described next.

Electrolytic Plating

In this process, a component is immersed in a conductive chemical solution (plating bath) that contains ions of the metal to be deposited. Electric current is passed through the component, which is the cathode (negative charge) in the electric circuit. The metal ions in the solution gain electrons at the surface of the component and attach themselves to the surface. A diagram of this process is shown in Figure 5.12. A wide variety of metals can be deposited using electrolytic plating, including copper, silver, gold, platinum, palladium, nickel, lead, tin, zinc, and chromium. Also, various alloys such as brass, bronze, tin-lead, and nickel iron can be deposited. Electrolytic plating can be used to deposit coatings onto most metals, and onto plastics and ceramics that have been properly prepared.

Conversion Coating

Conversion coatings are formed by immersing a metallic item into a chemical solution, with or without the application of electric current to the item. The reaction between the material and chemical solution results in the growth of a coating layer from the substrate surface. The coating is an integral part of the component surface and not just deposited over the surface. Thus, there is not a distinct mechanical interface between the coating and the underlying material.

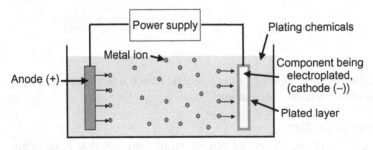

FIGURE 5.12

Diagram of the electrolytic plating process.

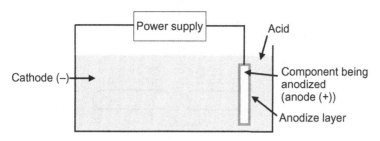

FIGURE 5.13

Schematic of the anodizing process.

Chemical conversion coatings include phosphate coatings formed on zinc or steel; chromate coatings formed on zinc, aluminum, cadmium, copper, magnesium, and silver; and oxide coatings formed on copper, iron, steel, and zinc alloys.

Electrochemical conversion coating, also referred to as anodizing, converts the surface of certain aluminum, magnesium, titanium, and zinc alloys to an oxide of the metal. This process involves immersion of a component in an acid solution and passing electricity through the component, with the component being the anode in the electrical circuit. This process is illustrated in Figure 5.13.

Painting

These processes involve application of a polymeric based liquid onto the surface of a component or subassembly, followed by heating to dry or cure the deposited material. Paint can be applied by methods such as spraying, dip coating, or curtain coating. Some of these processes are illustrated in Figure 5.14.

Powder Coating

This process involves spraying dry polymer powder particles onto the surface of the component, followed by heating to melt and flow the particles to form a nonporous coating. Prior to the heating step, the powder particles are held onto the surface of a component by electrostatic forces made possible by applying a voltage to the component during the coating process.

Physical Vapor Deposition

These processes take place in a vacuum chamber where metal atoms, which have been vaporized from a source by sputtering or evaporation, deposit on the surface of components. A schematic of this process is shown in Figure 5.15. Metals such as aluminum, copper, titanium, gold, platinum, chromium, and nickel can be deposited. Oxides, nitrides, and carbides can also be deposited by reacting vaporized and deposited metal with oxygen, nitrogen, or carbon containing gases during the deposition process.

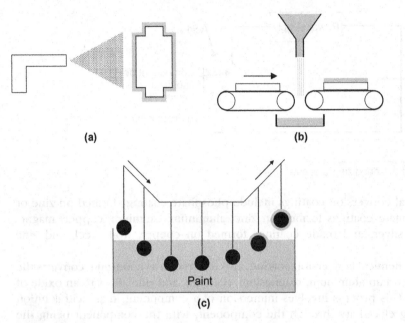

FIGURE 5.14

Schematics of painting processes. (a) Spray painting. (b) Curtain coating. (c) Dip painting.

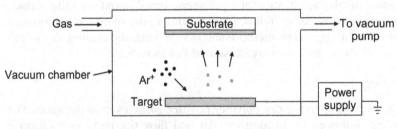

FIGURE 5.15

Schematic of a physical vapor deposition process.

Thermal Spraying

This process involves passing materials, in the form of wire, rod, or powder, through a spray unit where they are heated to a molten or near-molten state and form droplets. The droplets of material are accelerated in a gas stream and projected against the surface to be coated. On impact, the droplets flow into thin lamellar particles adhering to the surface, overlapping and interlocking as they solidify. The total coating thickness is usually generated in multiple passes of the coating device. The different thermal spray processes include flame spray, electric-

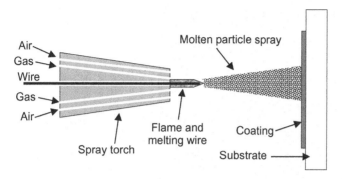

FIGURE 5.16

Schematic of a thermal spray process.

arc spray, plasma spray, high-velocity oxyfuel, and a detonation gun. A wide variety of metals, ceramics, and polymeric material can be deposited over various metal, ceramic, and polymeric components using thermal spraying. A schematic of a thermal spray process is shown in Figure 5.16.

Some components receive more than one secondary process treatment. For example, before painting, a surface may first be grit-blasted to clean and roughen the surface. Also, some materials are coated before forming the component (e.g., sheet steel, which is zinc or tin coated before the component is formed). The benefit is reduced cost compared to forming the component and then coating it afterward.

Instead of the bulk shaping processes, some components are fabricated using a series of deposition, etching, or chemical conversion processes to build the various structures of the component. A base material is used as a substrate on which all of the structures are formed. The substrate provides mechanical support and may also be incorporated into the functional structures of the component. Examples include components for electronics such as integrated circuits and microelectromechanical systems (MEMS). Examples of integrated circuit and MEMS devices are shown in Figure 5.17.

5.3 OVERVIEW OF JOINING PROCESSES

Joining processes are used to connect components and subassemblies. The broad categories of joining processes are as follows:

1. Welding
2. Soldering and brazing
3. Adhesive bonding
4. Mechanical fastening

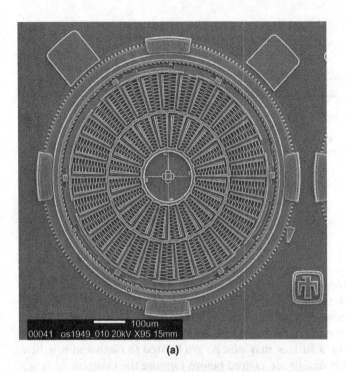

(a)

(b)

FIGURE 5.17

(a) MEMS torsional racheting actuator. (b) Integrated circuit. (*Source:* Part (a) courtesy of Sandia National Laboratories.)

These processes are described in more detail next. More information about joining processes can be found in *ASM Handbook,* Volume 6 (1993); Cubberly and Bakerjian (1989); DeGarmo et al. (2002); Easterling (1983); Humpston and Jacobson (2004); Kalpakjian and Schmid (2006); Messler (1993, 1999); and Swift and Booker (2003).

5.3.1 **Welding**

Welding is the process of forming a joint, either through localized melting (fusion) of the materials in the area of the joint or through high-temperature, solid-state reaction at the joint interface. Some of the fusion processes require the use of a filler material to fill the joint. Solid-state reactions take place below the melting point temperature of the materials being joined and involve the diffusion of the atoms of one material into the other or reactions to form compounds at the interface between the two materials being joined. The solid-state processes do not employ filler material. Some of the welding processes used are described next.

Arc Welding
Arc welding is used to join metals. An electric arc is created between an electrode and the components being joined in order to melt the components in the joint area. A filler material is fed into the joint at the same time. Arc-welding processes include shielded metal arc welding, metal inert-gas welding, tungsten inert-gas welding, plasma arc welding, and submerged arc welding. A diagram of the gas-tungsten arc welding process setup is shown in Figure 5.18. Examples of cross sections through arc welded joints are shown in Figure 5.19. Notice the distinct microstructures in and away from the area of each weld, and in the different materials.

Resistance Welding
Resistance welding involves passing an electric current through the components being joined to heat the interface enough to cause melting of the materials at the interface. No filler material is used. A clamping force, which is applied during the heating process, holds the components in place until the joint has solidified. Resistance welding processes are used to join metals and include spot welding, seam welding, flash welding, and projection welding. A diagram of the spot welding process setup is shown in Figure 5.20. An example of a spot-welded joint is shown in Figure 5.21, which shows a cross section of a spot weld between two pieces of 2024 aluminum sheet.

Laser and Electron Beam Welding
These processes involve the use of high-energy beams to weld metals. A clamping force, which is applied during the welding process, helps maintain intimate contact between the components being joined. No filler material is used. Examples of cross sections through an electron beam and laser beam welded joints are

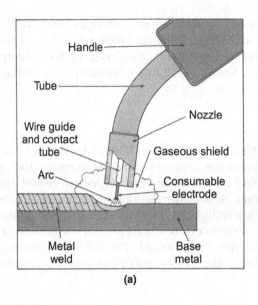

(a)

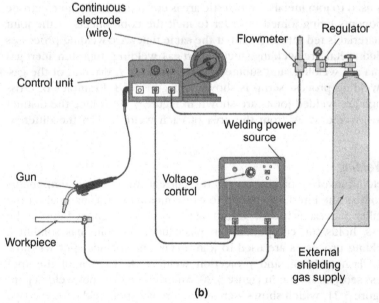

(b)

FIGURE 5.18

Schematics of the gas metal arc welding process setup. (*Source:* Beard, 1992. Reprinted with permission of AAVIM.)

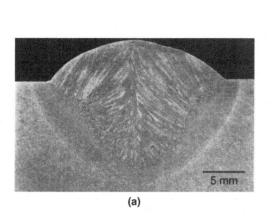
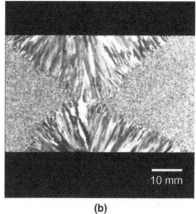

(a)   (b)

FIGURE 5.19

Examples of cross sections through arc welded joints. (a) Submerged arc-welded A-710 steel plate. (b) Tungsten arc-welded copper 11000 cold-rolled. (*Source: ASM Handbook, Volume 9, 1985. Reprinted with permission of ASM International.*)

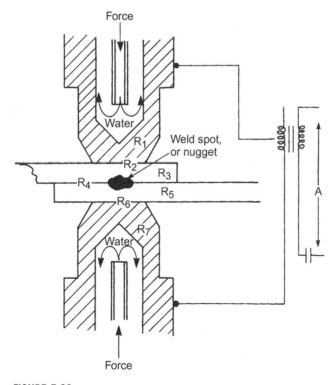

FIGURE 5.20

Diagram of the spot welding process setup.

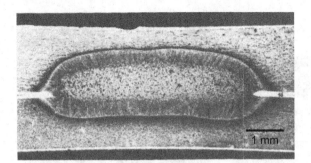

FIGURE 5.21

Cross section of spot welded 2024 aluminum sheet. (*Source: ASM Handbook,* Volume 9, 1985. Reprinted with permission of ASM International.)

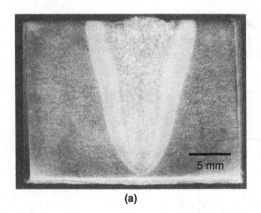

(a)

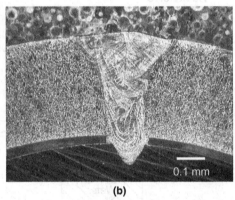

(b)

FIGURE 5.22

Examples of cross sections through (a) electron beam welded steel alloy plate and (b) laser beam welded Invar *(top)* to 304 stainless steel *(bottom)*. (*Source:* Part (a) from *ASM Handbook,* Volume 9, 1985. Reprinted with the permission of ASM International.)

shown in Figure 5.22. Figure 5.22(a) shows a cross section of an electron beam welded steel alloy plate. Figure 5.22(b) shows a cross section through laser beam welded Invar *(top)* to 304 stainless steel *(bottom)*.

Solid-State Welding

These processes are used to produce metal-metal, metal-ceramic, and ceramic-ceramic joints at temperatures below the melting point of the materials being joined, without the addition of filler metal. Either diffusion or deformation is used to form the joints. Solid-state welding processes include friction welding, diffusion

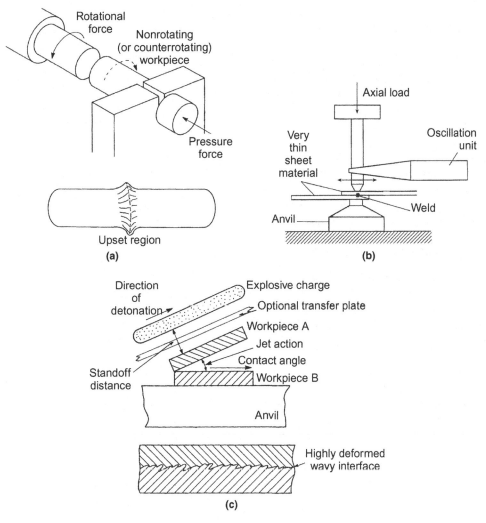

FIGURE 5.23

Schematics of solid state welding processes. (a) Spin. (b) Ultrasonic. (c) Explosion bonding. (*Source:* Parts (a) and (c) from Messler, 1993. Reprinted with the permission of Elsevier.)

bonding, and explosion bonding. Schematics of these processes are shown in Figure 5.23.

The heat for friction welding is produced by the relative motion of the two interfaces being joined. This method relies on the direct conversion of mechanical energy to thermal energy to form the weld, without the application of heat from another source. Under normal conditions, no melting occurs at the interface. Once the interface is heated, pressure is used to bring the weld interfaces into intimate contact. During this last stage, atomic diffusion occurs while the interfaces are in contact, allowing a bond to form between the two materials. Friction welding processes include stir and ultrasonic welding, which are used to join metals and join plastics. For plastics, the relative motion of the two interfaces being joined actually causes the plastics to melt at the bond line. When the melting occurs the motion is terminated and melted polymer at the joint interface is cooled under pressure. So, ultrasonic welding with plastics is not truly a solid-state welding process.

Solvent Bonding

Solvent bonding is used to bond thermoplastic materials. First, a solvent is applied to the surface of one of the components being joined, causing the material to soften. Then the two components are brought together and held in place with a clamping force until the interface is sufficiently set.

5.3.2 Soldering and Brazing

These processes involve applying a filler material to the components being joined, placing the components together, and then applying heat to cause the filler material to melt and react with the component materials. A metallurgical joint is formed between the parts. Soldering and brazing are used to join metal components and metal-coated ceramic components. Also, electrical components are soldered to circuit boards to make electronic circuits.

The major difference between brazing and soldering is the process temperature. The melting points of solders are less than 450°C, and the melting points of braze filler metals are greater than 450°C. The methods of applying heat to form solder joints include using solder irons, ovens, and resistance heating. Braze joints are formed by heating in a furnace, by induction heating, and by heating with a torch. Figure 5.24 shows examples of joints made by brazing. Figure 5.24(a) shows a cross section of two copper alloy tubes brazed together with a copper-silver-phosphorous alloy. The braze alloy is the material in the middle. Figure 5.24(b) shows a cross section of two aluminum alloy sheets brazed together with an aluminum-silicon alloy.

5.3.3 Adhesive Bonding

This process involves applying a polymeric material to one or both of the surfaces of the components being joined, pressing the components together, and allowing the adhesive to cure. Adhesives come in liquid, paste, and tape form. Liquids, pastes, and some tapes need to be cured after they are applied. Curing can take place at room temperature or may require heating in an oven. During the curing process, the adhesive material is converted from a relatively fluid, low-molecular-weight, non-load-bearing form to a solid, high-molecular-weight form able to bear loads. Epoxies, silicones, acrylics, and polyurethanes are used as adhesives. In some cases, a primer must be applied to the mating surfaces to enhance adhesion.

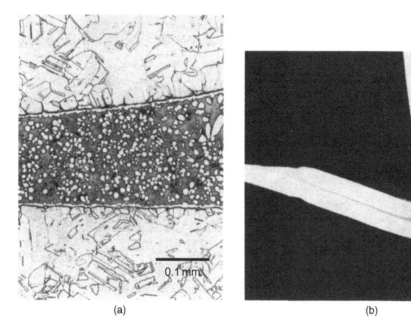

(a)   (b)

FIGURE 5.24

Examples of cross sections through brazed joints. (a) Copper alloy tubes brazed with a copper-silver-phosphorous braze alloy. (b) Aluminum alloy sheets brazed with an aluminum-silicon braze alloy.

5.3.4 **Mechanical Fastening**

Mechanical fastening includes joining components and subassemblies by press fitting, snap fitting, and using fasteners such as integral fasteners, threaded fasteners, and rivets. Integral fasteners are formed areas of a part that interfere or interlock with other parts of the assembly and include tabs, holes, embossed protrusions, and crimps. Threaded fasteners are separate parts that have internal or external threads and include bolts, nuts, and screws. Rivets, which consist of a head and body, are used to fasten two or more parts together by passing the body through a hole in each part and then forming a second head (clinching) on the body end. Examples of different methods of mechanical fastening are shown in Figure 5.25.

5.4 **OVERVIEW OF IN-PROCESS STRUCTURES**

In-process structures are formed by applying a material onto a subassembly. The material can be applied using a variety of methods including spraying, dispensing through a nozzle, and one of the coating processes discussed earlier. Most in-

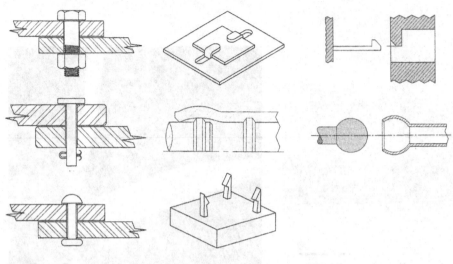

FIGURE 5.25

Examples of mechanical joints.

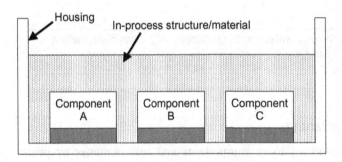

FIGURE 5.26

In-process structure schematic.

process materials are polymers and may require curing at an elevated temperature. A schematic of an in-process structure is shown in Figure 5.26. Figure 5.27 shows examples of in-process structures being formed. In Figure 5.27(a) a silicone coating is being sprayed over electronics. In Figure 5.27(b) a silicone gel is being dispensed into an electronics module.

5.5 PROCESS INPUTS AND OUTPUTS

Every manufacturing process consists of inputs and outputs. The inputs consist of the process variables, all of which influence the shape, size, material properties,

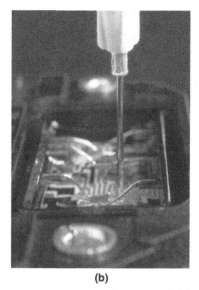

(a)   (b)

FIGURE 5.27

In-process structures being formed. (a) Silicone coating being applied over an electronics subassembly. (b) An encapsulating material being dispensed into an electronics unit. (*Source:* Part (a) reprinted with the permission of Asymtek. Part (b) reprinted with permission of Dow Corning Corporation.)

and material features of the product element being formed. The inputs also influence the type and number of defects present in a product element.

The process inputs consist of the following:

- Input materials
- Process equipment
- Process conditions
- Operator skill
- Manufacturing environment
- Data feedback from evaluations of the output

The inputs must be selected and controlled if the output is to meet its design requirements. Each process input is discussed in more detail in the following sections.

The output is the item produced, and it can be a material, component, subassembly, or assembly. Joints and in-process structures formed in subassemblies, and assemblies are also aspects of the output.

The relationship between the input variables, the process, and the output is shown in Figure 5.28. From the materials engineering perspective, there are two challenges to manufacturing, as follows.

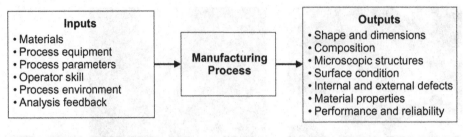

FIGURE 5.28

Process inputs and outputs.

1. Selecting process inputs that enable the formation of product elements with the required material features and material properties
2. Controlling the variation of the inputs such that the variation of the material features and material properties of the output is acceptable

5.5.1 Input Materials

Input materials are used to fabricate components and to form joints and in-process structures. They must have materials, features, and properties that enable the formation of product elements that meet their design requirements. Furthermore, their materials features and properties should enable the product elements to be easily formed.

Two categories of input materials are defined: primary input materials and secondary input materials. *Primary input materials* are used to fabricate components, fill joints, and form in-process structures. Primary input materials come in various forms, such as sheet, bar, slab, wire, liquid, paste, pellet, powder, and tape, and are fabricated using a variety of processes. The properties, composition, and microstructure of these materials depend on the manufacturing processes used to create the materials. Examples of primary input materials and their uses are as follows:

- Metal sheet for fabricating stamped components
- Metal bar for fabricating machined components
- Plastic pellets for fabricating injection molded components
- Ceramic powder for fabricating ceramic components
- Porcelain enamel for coating steel and iron components
- Adhesives, solder pastes, and braze compounds used to join components

Some of these materials are shown in Figure 5.29.

Secondary input materials are materials that make up a component-in-process that will go through a secondary process or allow components to be joined or incorporated into an in-process structure. Components to be joined can be individual or part of a subassembly. An example of components used as input materials

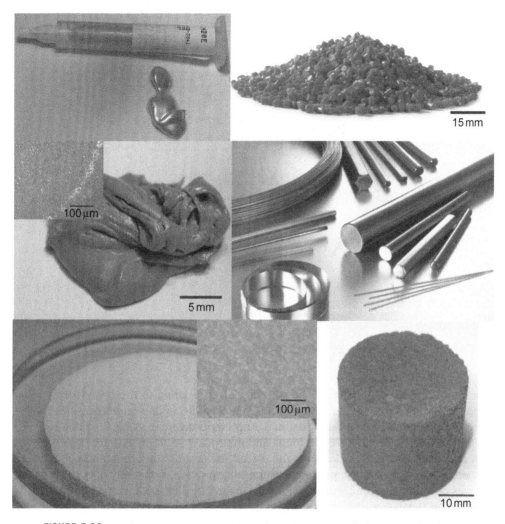

FIGURE 5.29

Examples of primary input materials used to fabricate components, form joints, and form in-process structures. From upper left corner and moving clockwise, epoxy adhesive, plastic pellets for injection molding, metal bar, sheet and wire, plastic bulk molding compound, ceramic powder, solder paste.

is two components to be joined by soldering, brazing, welding, or adhesive. The features and properties of the components' materials and surfaces in the area to be incorporated into the joint influence the ease of forming the joint as well as its features and properties. Another example is a component or set of components (i.e., subassembly) that is to be coated. The features and properties of a compo-

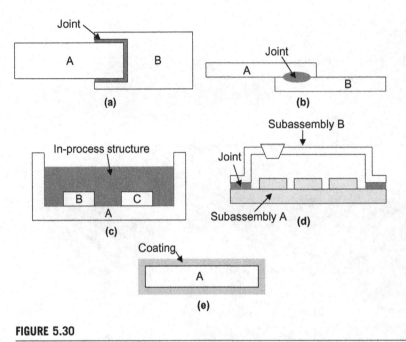

FIGURE 5.30

Schematic of secondary input materials.

nent's surface have an impact on the adhesion of the coating to the component. Figure 5.30 shows some schematics of components as secondary input materials, where the materials and surfaces that make up components A, B, or C, and subassemblies A and B, are involved in joints and in-process structures or are coated.

A manufacturing process involves the use of one or more input materials. Fabrication of a component by molding, casting, deformation, and material removal involves one primary input material. Powder material processes involve both the powder and the gases used to form the atmosphere for sintering. Processes such as those for surface modification, coating, joining, and forming an in-process structure involve at least one primary input material (e.g., gases, paint, joint filler, in-process material) and one or more secondary input materials.

Input Material Processing Capabilities

An input material has certain capabilities with respect to the ability to manipulate it to form a product element that has the desired shape, dimensions, material properties, material features, and types and number of defects. These capabilities can be appraised based on the following two criteria:

1. *Composition and microstructure.* An input material must have the composition and microstructure that enable the formation of a product element that satisfies all of its design requirements. For example, to make a ceramic com-

ponent that is 99% alumina with 1% silica at the grain boundaries, the ceramic powder must consist of 99% alumina particles and 1% silica particles. A metal bar must have a certain alloy composition and microstructure in order to make a component, using material removal processes, that has the desired composition and microstructure.

2. *Ease of use.* An input material should favorably interact with the other process inputs to allow a product element to be formed without difficulty. Examples are plastic pellets with additives that allow for easy removal of an injection-molded component from the mold, the ability of a solder paste to clean the surfaces being joined, the ease with which a metal component is machined from a bar of material, and the manner in which an adhesive spreads over surfaces to be joined. How an input material interacts with the other process inputs depends on its material features and properties, and affects the ability to form a product element that satisfies all of its design requirements.

The processing capabilities of an input material depend on its material features and properties, as well as the defects present within the material. The required capabilities for an input material for a specific process depend on the particular input material, the requirements of the process output, and the other process inputs. Consider two different manufacturing lines with similar processes and outputs, but with different equipment and process conditions. A material with capabilities that are acceptable for one manufacturing line may not be acceptable for the other. The capabilities of an input material used to make a low-performance, low-reliability item may not be acceptable for making a high-performance, high-reliability item.

The significant features, properties, and defects for a few input materials are as follows.

Ceramic Powder

Ceramic powder consists of ceramic particles and additives that improve a powder's ease of use during component fabrication. Additives include a binding agent to hold the powder together after compaction and a release agent to enable a compacted component to be easily removed from the compaction die. The significant features and properties of ceramic powder are the particle size and its size distribution, the particle composition, the amount of die release agent and its composition, the amount of binder and its composition, and the type and amount of contamination. Defects in the powder include clumps of ceramic particles and foreign particles.

Solder Paste

Solder paste is used to join electrical components to printed circuit boards to create an electronic circuit. An example of solder joints was shown in Chapter 2 for the cordless telephone. Solder paste is smeared, by hand or robot, over the surface of

FIGURE 5.31

Electrical components soldered to a circuit board.

a stencil that has been placed over a circuit board. The solder paste is pushed through holes in the stencil onto metal pads on the circuit board. The stencil is removed, and electrical components are placed on the circuit board, with their leads in contact with the solder paste. The circuit board is then sent through an oven, where the circuit board is heated until the metal in the solder paste melts and forms a metallurgical joint between the electrical component leads and the pads on the circuit board. A complete electronic circuit is shown in Figure 5.31. A schematic of the solder paste application and reflow process is shown in Figure 5.32. Solder paste consists of metal particles, flux, and vehicle.

During the soldering process the metal particles melt and form metallurgical bonds between the components being joined. The composition of the metal particles controls the melting point. The size of the powder particles influences the melting behavior of the paste and some of the paste-spreading and stencil-release properties. The flux cleans the surfaces to be joined and the solder powder particles before the solder particles melt. This cleaning enables the metallurgical reaction to occur between the paste and each of the components being joined.

The chemistry of the flux affects its ability to clean the metal surfaces being soldered. The vehicle provides a means for suspending the metal powder in the paste and gives the paste certain characteristics with regard to the way it spreads on the stencil, the way it releases from the stencil when the stencil is pulled away from the circuit board, and the shape that the paste holds on the circuit board as the board is heated in the oven.

The significant features of solder paste are the metal particle size and size distribution, metal particle composition, flux composition, and vehicle composition. The significant properties are paste viscosity and paste thixotropy (response to

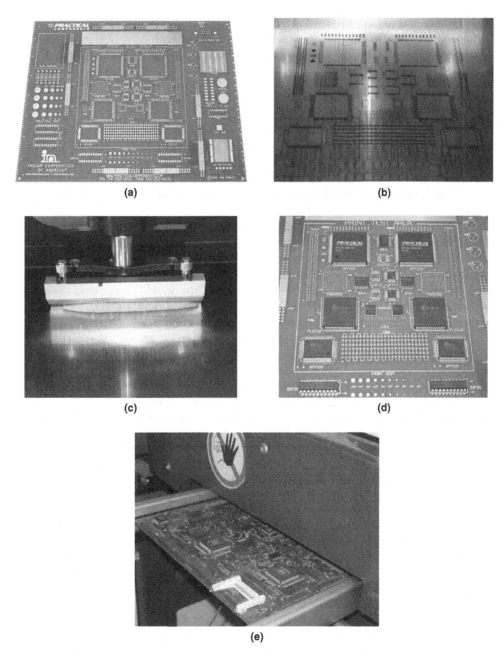

FIGURE 5.32

Solder paste application and reflow process. (a) Circuit board. (b) Stencil. (c) Solder paste being spread over the stencil. (d) Electrical components placed on solder paste on the circuit board. (e) Circuit board with components going through an oven. (*Source:* Parts (a)–(d) courtesy of The Indium Corporation.)

shear forces). The required control of these characteristics increases as the spacing between the leads of electrical components decreases and with increased variation of the quality of the circuit board metal pads and electrical component metal leads.

Thermoplastic Pellets

Thermoplastic pellets for injection molding consist of thermoplastic and possibly other materials such as inorganic mineral fillers, reinforcement material such as glass fiber or glass particles, mold release (which allows the component to detach more easily from the mold), chemical flame retardant, heat stabilizers (which prevent the plastic from degrading during processing or high temperature use), and colorant.

The flow characteristics of the plastic into the mold depend on the following: (1) the molecular weight and molecular weight distribution of the plastic, (2) the amount of filler and reinforcement materials, (3) the physical size of filler and reinforcement materials, (4) how well the filler and reinforcement materials are distributed throughout each pellet, and (5) the design of the mold. The plastic flow characteristics into the mold affect the density of the component, the distribution of filler in the component, and the presence of defects such as knit lines.

The significant material features for thermoplastic pellets are the plastic composition, plastic average molecular weight and molecular weight distribution, mineral filler composition, mineral filler particle size, mineral filler particle size distribution, and composition and concentration of the chemical additives. Significant properties of the pellets are the glass transition temperature, melting temperature, bulk density, sensitivity of the molten material to shear, and melt flow rate.

Metal Components to Be Resistance Welded

Forming a good resistance weld joint between two metal components depends on the composition of the components and the cleanliness and roughness of their surfaces. The metals' compositions affect their melting point temperatures, which affect the ability to get good mixing between the materials when the welding current is applied. Contamination on the surface interferes with good contact between the materials and can also get included in the weld, reducing its strength. The surface roughness affects the ability to get good contact between the components before forming the weld. The weld spot will be smaller than expected if the components are rougher than expected.

Base Material to Be Coated

The adhesion of a coating to an underlying material depends on the surface condition of the base material. The presence of surface oxidation prevents proper adhesion for some coatings. The presence of surface contamination such as dirt or oil causes poor adhesion. The texture of the substrate surface is important for the adhesion of coatings that depend on mechanical bonding to the surface. For these situations, the substrates are roughened so that the coating can "grab" onto the surface.

In addition to considering the capabilities of a material, design and manufacturing teams must also consider the variation of the material features and properties of an input material. There will be a certain amount of variation from batch to batch, which means that the process capabilities of the input material will fluctuate as well. The amount of variation depends on the particular material, the method of producing it, and a supplier's manufacturing skill.

5.5.2 Process Equipment

For any given manufacturing process there is a wide selection of equipment manufacturers, grades of equipment, and ages of equipment. The different pieces of equipment will have different capabilities and behaviors, which will affect a manufacturing team's ability to control process variations.

The important factors for a piece of equipment are its process controls, fixtures, and stability.

Process Controls

During processing, the input materials are exposed to various mechanical, electromagnetic, thermal, chemical, electrochemical, and radiation conditions. The process controls are used to set and monitor the process conditions. The conditions that can be controlled and the precision with which adjustments can be made vary from equipment to equipment. Even though many processes have several conditions that can be controlled, the need to control certain process conditions will depend on the acceptable levels of variation of the material properties, material features, size, and shape of the product element being formed. Furthermore, some process conditions may have greater impact on the attributes of the output compared to the other conditions. The need to control the less significant conditions will be greater when forming product elements that cannot tolerate large variations of certain attributes compared to less demanding product elements. Finally, the required accuracy and precision of the controls increases as the tolerance of output variation decreases.

Consider skillets of different quality that have nonstick coatings applied by spray painting the coating material onto the pan base metal. A high-quality skillet manufacturer will control the spray pressure and spray time to deposit a coating of consistent thickness from skillet to skillet. A lower-quality skillet manufacture may control neither, and just spray the coating until it looks thick enough.

Fixtures

A fixture holds a work piece in place during processing or is used to perform a process step on a work piece. A work piece is the item being processed into a component, subassembly, or assembly. Examples of fixtures include molds for injection molding and die casting and electrodes for resistance welding. The significant factors are the fixture's materials and its physical construction.

The materials that make up a fixture must be selected so that they do not quickly degrade during use. Degradation can take on such forms as cracking, wear, and corrosion. The degradation results in changes in the work piece placement or the manner in which the process operates on the work piece. For example, the wear of resistance welding electrodes will result in reduced force applied to the work piece and change the contact interface between the electrodes and the work piece. These changes will lead to changes in the microstructure of the weld joint, the joint strength, and the defects present in the joint. Periodic fixture maintenance may be required to address any normal deterioration before it becomes a problem.

Stability

Equipment stability refers to the ability of a piece of equipment to maintain the process conditions at the desired setting. Drift of the values of the process conditions from their settings will result in variation of the process output. As the stability decreases, the amount of drift increases. Drift occurs because of variations in the mechanics and electronics of the equipment, which can arise as a result of poor equipment design, equipment component degradation (e.g., wear and corrosion), and fluctuations in a factory's environment (e.g., temperature or humidity). Poor design results in a sloppy fit between components or the use of inferior components that do not have stable performance. Degradation can occur in both mechanical and electrical components. For example, wear of electrical contacts results in intermittent electrical contact, which results in fluctuations of the electrical control signals. Fluctuations in factory temperature will cause the components to expand and contract, which can affect the manner in which they move during operation.

5.5.3 Process Conditions

The process conditions are the mechanical, electromagnetic, thermal, chemical, electrochemical, and radiation means used to manipulate the input materials during processing. Examples of the process conditions for different manufacturing processes are listed in Table 5.1.

The process conditions are adjusted using the process equipment controls. In some cases, the controls adjust the process conditions in a direct manner. For example, the temperature in an oven is adjusted via a temperature controller. The screw speed in an injection molder is adjusted via a speed controller. In other cases, the controls adjust the process conditions in an indirect manner. For example, the composition of a gas atmosphere inside of an oven may be adjusted by controlling the line pressure of the various gases flowing into the oven. The deposition rate of

Table 5.1 Input Process Parameters for Different Processes

Process	Process Condition
Ceramic sintering	Rate of heating, sintering temperature, sintering time, sintering atmosphere composition
Plastic injection molding	Screw speed, barrel temperature, injection pressure, mold temperature
Metal forging	Work piece temperature, hammer force and speed
Adhesive joining	Adhesive dispense pressure, dispense time, cure temperature, cure time

an electroplated coating is controlled by adjusting the electric current flowing between the component being coated (the cathode) and the anode.

The *process parameters* (i.e., temperature, screw speed, line pressure, current, and belt speed) are the variables controlled to obtain the desired process conditions. A process parameter *set point* is the specific setting of a process parameter.

Consider the process conditions for the fabrication of a ceramic component by compaction and sintering. The compaction pressure affects how tightly the powder particles are pushed together, which affects the porosity and density of the sintered component. The rate of heating up to the sintering temperature must be properly controlled so that when the component is heated it does not crack because of thermal stresses. The maximum temperature affects the amount of porosity and the degree of grain growth in a sintered component. Furthermore, too high of a temperature allows the component to sag under its own weight. The sintering time is important for achieving the desired microstructure. If the time at the maximum temperature is too long, then the component will be too dense or have excessive grain growth. If the time is too short, then there will be excessive porosity and the grain structure will not properly evolve. Finally, for certain materials the atmosphere in which the sintering occurs must be of a certain composition (e.g.. air, vacuum, nitrogen) to keep the component from oxidizing.

The design team must understand the effects of the process conditions on the product elements if the product is to consistently meet its performance and reliability requirements. This includes understanding the effects of individual conditions, the interactions between different process conditions, and the variation of the conditions on the process output. Some experimentation is often required to obtain this information. The amount of experimentation and the costs and time needed to conduct the experiments increase as the complexity of the process and its output increase. Design of experiments is an engineering tool that allows engineering teams to methodically characterize the effects of controlled variations of the process conditions and input materials on the output variables; for examples, see Mason (2003), Montgomery (2004), and Ross (1995).

As with the input materials, the significance of any single process condition depends on the specific process and the requirements of the process output. If the output has few requirements and none of them are stringent, then it may not be necessary to exert tight control over the conditions. Conversely, if the output has many requirements that must be tightly controlled, then close control of the conditions is required.

5.5.4 Operator Skill

Operators are the workers on a manufacturing production line. They are responsible for performing a variety of tasks associated with the manufacturing process, such as handling the input materials and the item being produced, operating the process equipment, adjusting the process parameters, and evaluating the process output. For example, a manufacturing production line for welding two components together might involve placing the components in a welding fixture, setting the equipment controls to obtain the desired welding conditions, welding the components, removing the components from the welding fixture, and evaluating the weld joint.

Performing these tasks requires a certain amount of knowledge and skill. Differences in the knowledge and skill from operator to operator cause variations in the process output because there will be differences in how a piece of equipment is used and in the adjustments to the process parameter set points. The amount of operator skill required increases as the number of process steps increases, the complexity of each process step grows, and the amount of acceptable variation of how a process step is performed decreases.

Production lines that have partial or complete process automation obviously require less operator involvement compared to nonautomated lines. The benefit of the automation is reduced variation. However, good control of the automation is also required. Otherwise, a great number of defective products may be manufactured before anyone notices.

5.5.5 Manufacturing Environment

The manufacturing environment in a factory can be hot, loud, and humid. The operation of equipment or passing traffic may cause the floor to vibrate. Such conditions can cause the process equipment to behave erratically, affecting the process output. The degree to which the manufacturing environment needs to be controlled depends on the sensitivity of the process inputs to the environmental conditions in a factory. In some factories, the temperature and humidity are monitored and controlled because the processes are sensitive to fluctuations.

5.5.6 Analysis Feedback

Most manufacturing processes have a feedback mechanism for making adjustments to the process inputs based on data from evaluations of the process output. The evaluations are performed to verify that the output meets its requirements, and are used to assess and quantify performance, dimensions, shape, composition, microstructure, material properties, and defects.

The process inputs are adjusted based on the data from the evaluations. For example, if the thickness of a coating on a component is measured and found to be too thick or thin, then an adjustment may be made to the setting for one of the process conditions in order to increase or decrease the thickness.

Because process adjustments are based on evaluation data, the quality of the data is very important. If the data are unreliable, then unnecessary adjustments will be made or necessary adjustments will not be made. In either case, the team will not have an accurate understanding of the attributes of the process output. The team may believe that the process output meets its design requirements when actually it does not.

The quality of the data from an output evaluation depends on the measurement system used to perform the evaluation. A measurement system includes the equipment, methods, and personnel used to perform an evaluation. It is subject to the same considerations and concerns discussed here for manufacturing processes. The quality of data evaluation is a function of the accuracy of the data over time and the precision of the data, both of which can be measured. Information about developing reliable measurement systems can be found in *Measurement Systems Analysis* (2002).

5.5.7 Process Output

The process output variables consist of the characteristics of the product element, subassembly, and assembly produced. The characteristics of a product element are its performance, material features, material properties, size, shape, and manufacturing defects. The characteristics of a subassembly and assembly consist of their size, shape, and performance attributes.

There is a range of acceptable values for any particular characteristic. The range can be represented by upper and lower limits, or have a maximum or minimum value.

The types of defects that can form in a product element depend on the type of manufacturing process used to create it. Some of the defects that can result from different processes are as follows:

Ceramic sintering. Cracks, agglomerates (clumps of particles that did not sinter), large pores, inclusions (contaminant particles), excessively large grains, excessive porosity inside grains, unwanted glass phase at grain boundaries.

Arc welding. Cracks, porosity, nonmetallic inclusions, incomplete penetration, lack of fusion of the components being welded.

Plastic injection molding. Voids, splay on surface (whitening), surface bubbles, knit lines, blisters, burn marks, oxidation, embedded contaminants (from operating environment or packaging material), discoloration, excessive shrinkage or warpage, sink marks, flow marks.

Metal die casting. Cold shuts (seams that form when two streams of molten metal meet but do not completely merge; caused when molten metal is poured at too low a temperature or run over too long a distance), galling or drag marks on the surface, poor fluidity, die soldering, shrinkage porosity, hot cracking, gas porosity, heat check fins.

Metal forging. Cracks, laps, seams.

Metal sheet forming. Tears, wrinkles, cracks, orange peel.

Powder coating. Pinholes, bubbles.

Electroplated metal. Burned metal, pores, delamination, blisters.

The number and physical size of defects present in a product element depend on the skill with which the process inputs are selected and controlled.

5.6 PROCESS VARIATION AND CAPABILITY

For any process there will be variation of the process input variables, which will result in variation of the output characteristics. The input variation is acceptable if the resulting output meets its design requirements. Most processes can tolerate a certain amount of variation of the inputs. The larger the input variation that can be tolerated the better because less control of the process inputs is required. However, if the variation of any of the output characteristics is excessive, the input variation related to the output variation must be reduced.

The variation of input variables can be placed in one of two general categories: common cause and special cause.

Common cause variations. These are random fluctuations of the process inputs that are predictable and always present. Therefore, common cause variations result in output variation that is stable and repeatable over time. Examples are the random variations of the particle size of a ceramic powder between 10 and 20 μm diameter or the random variations of the temperature of an oven between 298°C and 302°C.

Special cause variations. These are fluctuations that are intermittent and unpredictable. Consequently, they result in output variations that are unstable and unrepeatable over time. Examples are equipment malfunctions, a batch of input material that does not meet its requirements, and an operator setting a process parameter to an incorrect set point on a piece of equipment.

A process capable of consistently producing output that meets its design requirements necessitates reducing common cause variations to acceptable levels and eliminating special cause variations. A process is said to be *in control* once special cause variations have been eliminated. A process is said to be in control and *capable* when the special cause variations have been eliminated and common cause variations have been reduced so that only an insignificant portion of the output fails to meet design requirements. More information about process control, capability, and process variation analysis can be found in *Statistical Process Control* (2005).

An output attribute's sensitivity to common cause variations depends on the design of the process output and the manufacturing process used. If the requirements of the output are too narrow compared to the common cause variations of the manufacturing process, then it will not be possible to develop a capable process. For this reason, it is important for a design team to understand the capabilities of a manufacturing process when making design decisions.

Developing a capable process requires understanding the effects of variations of the input variables on the significant output characteristics and then controlling for that variation as required. Not all input variables have the same impact on the output characteristics. For some input variables, relatively large variations are acceptable; for others, only relatively small variations can be tolerated. More control over the variation is required for the latter case compared to the former case.

Frequency curves are used to quantify the amount of variation of an input variable or output characteristics and to explain the effects of process input variations. A *frequency curve* plots the probability of the occurrence of a particular value of an attribute over a range of values. The attributes of interest are the characteristics of the input and output variables. Examples of attributes that can be plotted are ceramic powder particle size, sintering furnace temperature, paint thickness, deposition rate, and joint strength.

The shape of a frequency curve depends on the data that it represents. For most processes, the variation of the process inputs and the process output characteristics can be described by a *normal frequency curve,* which is associated with a *normal frequency distribution.* An example of a normal frequency curve is shown in Figure 5.33. The value of the attribute of interest is shown on the x-axis. The probability of the occurrence of a particular value is shown on the y-axis. In Figure 5.33, the resistance of electrical resistors produced on a production line is shown on the x-axis. For this distribution, the probability of producing a resistor with a resistance of 50 ohms is about 8%.

A distribution is described by two parameters: the sample *mean* $\bar{x}$ and the sample *standard deviation* s. The mean is the average value of the samples taken and is equal to the sum of all of the values of the samples measured, divided by the total number of samples.

$$\bar{x} = \frac{v_1 + v_2 + v_3 + \ldots + v_N}{N}$$

where v_N is the value of a characteristic or parameter for the Nth sample and N is the total number of samples. The sample standard deviation is a measure of the spread of the data around the mean and is calculated according to the following equation:

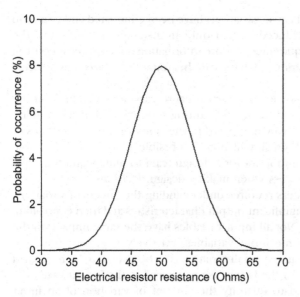

FIGURE 5.33

Example of a normal frequency curve.

$$s = \sqrt{\frac{(v_1 - \bar{x})^2 + (v_2 - \bar{x})^2 + (v_3 - \bar{x})^2 + \ldots + (v_N - \bar{x})^2}{N-1}}$$

A small sample standard deviation indicates close sample data clustering about the mean. A large sample standard deviation indicates wide scattering of the sample data about the mean. As the magnitude of the sample data variation increases, the standard deviation increases, and the spread of the distribution increases.

For any normal distribution, 68.2% of the samples evaluated will have a value within $\pm 1\sigma$ of the mean. This is equal to the area under the frequency curve within $\pm 1\sigma$ of the mean and is indicated by the shaded area in Figure 5.34; 95.4% and 99.7% of the samples evaluated will have a value within $\pm 2\sigma$ and $\pm 3\sigma$ of the mean, respectively.

Figure 5.35 shows three normal frequency curves. The mean for distribution C is greater than the mean for distribution A. The standard deviations for distributions A and C are the same, so the spread in the data is the same. Distributions A and B have the same mean, but the standard deviation for distribution B is greater than that for distribution A. Thus, there is more variation in the data for distribution B compared to distribution A. Finally, the mean is lower and the standard deviation higher for distribution B compared to distribution C.

The concept of process capability is illustrated in Figure 5.36. Consider two production lines, A and B, which both make the same product for which the

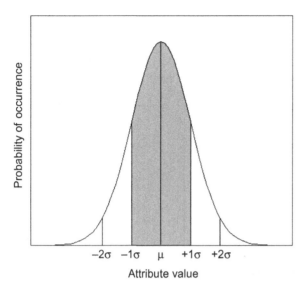

FIGURE 5.34

Standard deviation for a normal distribution frequency curve.

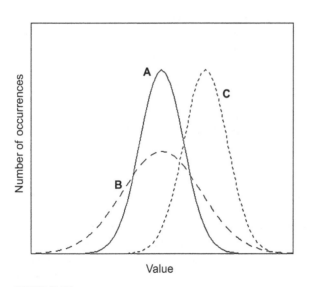

FIGURE 5.35

Three different normal distribution frequency curves.

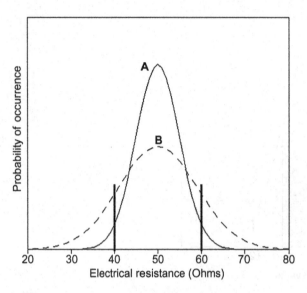

FIGURE 5.36

Illustration of process capability.

electrical resistance must be 50 ± 10 ohms. The input materials, equipment performance, process conditions, operator skill, measurement systems, and manufacturing environment result in curves A and B for the electrical resistance of the
output from production lines A and B. Both output distributions have the same
mean, but the variation in B is much greater than in A. This is due to more variation of one or more of the input variables for production line B compared to production line A.

The overall impact of the differences in the distributions of the electrical resistance of the output is that a significantly larger percentage of the product from
process B does not meet the design requirements compared to the product from
process A. Thus, process A is more capable than process B. Consequently, the
amount of scrap product from production line B is higher than that from production line A.

If products from both production lines are sold at the same price, then the
profits for production line B will be lower than those for production line A because
of the extra costs associated with making an unsellable product on line B. To make
the same profits, the sellable product from line B will have to be sold at a higher
price than the comparable product from line A. Both cases assume identical manufacturing costs for lines A and B.

There are three ways to improve the capability of a process. The first is to
increase the range of acceptable values of the output characteristics. This a
product design decision, which may be difficult to make after a product goes into

production because the requirements were selected based on customer wants and needs. Changing any output acceptance requirements reduce the ability to satisfy the wants and needs of the intended customers.

The second way to improve the capability of a process is to reduce the variation of the input variables. As mentioned earlier, for any particular process, the variation of some input variables has less impact on the output variation than that of other input variables. It is up to manufacturing teams to determine the significant input variables and then understand how their variation can be reduced.

The third way to improve process capability is to design product elements, subassemblies, and assemblies so that their design requirements are within the expected variation of processes to be used. This approach involves considerations about design for manufacturing (Swift and Booker, 2003; Booker, Swift, and Raines, 2001; Nevins and Whitney, 1989), which help design teams make decisions that result in capable manufacturing processes. This approach is in contrast to the approach of designing a product and requiring manufacturing engineers to figure out how to make it.

Understanding the relationship between the process inputs and the output characteristics is crucial for making components, subassemblies, and assemblies that consistently meet their performance and reliability requirements. This understanding requires characterization of the input variables and their impact on the output characteristics, and it is conducted during process development and efforts to improve manufacturing yields. This is discussed in more detail in Chapters 10 and 11.

REFERENCES

ASM Handbook, Volume 14a: Metal Working: Bulk Forming, Tenth Edition, ASM International, 2005.

ASM Handbook, Volume 14b: Metal Working: Sheet Forming, Tenth Edition, ASM International, 2006.

ASM Handbook, Volume 15: Casting, ASM International, Tenth Edition, 1988.

ASM Handbook, Volume 16: Machining, ASM International, Tenth Edition, 1989.

ASM Handbook, Volume 4: Heat Treatment, Tenth Edition, ASM International, 1991.

ASM Handbook, Volume 5: Surface Engineering, Tenth Edition, ASM International, 1990.

ASM Handbook, Volume 6: Welding, Brazing, and Soldering, ASM International, 1993.

ASM Handbook, Volume 7: Powder Metal Technologies and Applications, Tenth Edition, ASM International, 1998.

ASM Handbook, Volume 9: Metallography and Microstructures, Ninth Edition, ASM International, 2004.

ASM Handbook, Volume 9: Metallography and Microstructures, Ninth Edition, ASM International, 1985.

Berins, M., *Plastics Engineering Handbook of the Society of the Plastics Industry*, Fifth Edition, Van Nostrand Reinhold, 1991.

Booker, J.D., M. Raines, K.G. Swift, *Designing Capable and Reliable Products*, Butterworth-Heinemann, 2001.

Cubberly, W.H., and R. Bakerjian, editors, *Tool and Manufacturing Engineers Handbook Desk Edition*, Society of Manufacturing Engineers, 1989.

DeGarmo, E.P., J.T. Black, and R.A. Kohser, *Materials and Processes in Manufacturing*, Ninth Edition, Wiley, 2002.

Easterling, K., *Introduction to the Physical Metallurgy of Welding*, Butterworths and Co., 1983.

Humpston, G., and D.M. Jacobson, *Principles of Soldering*, ASM International, 2004.

Kalpakjian, S., and S.R. Schmid, *Manufacturing Engineering and Technology*, Fifth Edition, Pearson Education, 2006.

Mason, R.L., R.F. Gunst, and J.L. Hess, *Statistical Design and Analysis of Experiments, with Applications to Engineering and Science*, Second Edition, Wiley-Interscience, 2003.

Measurement Systems Analysis, Third Edition, Automotive Industry Action Group, 2002.

Messler, R.W. Jr., *Joining of Advanced Materials*, Butterworth-Heinemann, 1993

Messler, R.W. Jr., *Principles of Welding: Processes, Physics, Chemistry, and Metallurgy*, John Wiley and Sons, 1999.

Montgomery , D.C., *Design and Analysis of Experiments*, Sixth Edition, Wiley, 2004.

Nevins, J.L., and D.E. Whitney, *Concurrent Design of Products and Processes: A Strategy for the Next Generation in Manufacturing*, McGraw-Hill, 1989.

Reed, J.S., *Principles of Ceramic Processing*, Second Edition, Wiley-Interscience, 1995.

Ross, P.J., *Taguchi Techniques for Quality Engineering*, Second Edition, McGraw-Hill Professional, 1995.

Richerson, D.W., *Modern Ceramic Engineering: Properties, Processing and Use in Design*, Third Edition, CRC Press, 2006.

Rosato, D., and D. Rosato, *Injection Molding Handbook*, Second Edition, Chapman & Hall, 1995.

Statistical Process Control, Second Edition, Automotive Industry Action Group, 2005.

Swift, K.G., and J.D. Booker, *Process Selection: From Design to Manufacture*, Second Edition, Butterworth-Heinemann, 2003.

Zhang H., and J. Senkara, *Resistance Welding: Fundamentals and Applications*, Taylor and Francis, 2006.

Degradation and Reliability of Materials

6.1 INTRODUCTION

The materials in a product will often degrade during use. Metallic components corrode when exposed to saltwater, plastics become brittle and their colors fade when exposed to sunlight, and the coatings on surfaces wear away. As a material degrades, the ability of the affected product element to satisfy its performance requirements decreases. If there is enough degradation, the product element will stop functioning altogether.

A product's reliability is based on the reliability of the materials used within it. The reliability of a material depends on the manner in which it degrades and the degradation rate in response to the mechanical, electromagnetic, thermal, chemical, biological, electrochemical, or radiation conditions to which it is exposed. In some cases, the conditions have no impact on a material, and the material is able to perform as required over the expected life of the product. In other cases, the conditions cause a material to degrade, but the material is still able to perform as required over the expected life of the product. In a third case, a material degrades so much that the product element in which the material is used can no longer perform as required. At this point the product element fails, resulting in the loss of one or more of the product's functions. This is a problem if the failure occurs during the part of the product's life when it is expected to perform without failure.

The failure of a product because of the degradation of its materials can take on one of the following two forms:

1. A particular product functionality slowly degrades below the minimum requirements. This occurs when the materials that make up a product element slowly degrade, accompanied by changes in the materials' properties. The product still performs, but not as well as it did before the product element degraded below the minimum requirements.

2. A particular functionality is completely lost without warning. In some cases, this loss leads to unexpected behavior that can cause damage or personal injury.

An example of the first situation is a windshield wiper blade insert that slowly hardens and loses flexibility with continued exposure to sunlight, heat, and chemicals (e.g., windshield wiper fluid, saltwater). This degradation of the wiper insert's material properties leads to a slow deterioration in its ability to wipe water cleanly away from the windshield as it no longer conforms perfectly to the shape of the windshield. As the insert material continues to degrade, the driver's ability to see through the windshield decreases. The owner of the automobile may not notice the degradation in performance at the onset. However, after the windshield wiping performance has degraded sufficiently, the owner will notice.

An example of an abrupt loss of functionality is a motor shaft in which a crack starts and grows during use, as shown in Figure 6.1. The cross-section of the shaft bearing the load decreases as the crack grows. As a result, the actual stress on the load-bearing portion of the shaft increases. The shaft fractures when the actual stress increases to the point where it is greater than the material can bear.

There are many degradation mechanisms that can occur in materials (e.g., composition change, loss of material, embrittlement, phase changes, or cracking). The occurrence of any of them can result in the degradation of a material's properties, leading to outcomes such as loss of strength, reduced electrical or thermal conductivity, changes in dielectric strength, increased contact resistance, and increased friction.

Selecting materials that enable a product element to satisfy its reliability requirements necessitates that design teams understand the materials' response to the conditions encountered during use. This chapter discusses the mechanisms of material degradation and methods for evaluating the degradation and reliability of materials. The first part of this chapter discusses materials reliability and degradation in general and some of the modes of materials degradation and failure. The second part discusses the tests that can be performed to evaluate the degradation and reliability of materials.

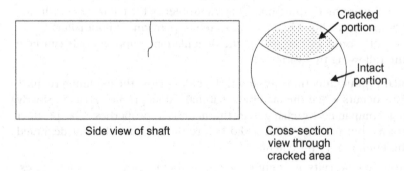

Side view of shaft   Cross-section
                           view through
                           cracked area

FIGURE 6.1

A schematic of a crack propagating through a motor shaft.

6.2 MODES OF MATERIAL DEGRADATION AND FAILURE

Some of the modes of materials degradation are explored next. The discussions are intended to provide a general sense of the various modes of degradation that occur in materials. Not all general modes are discussed, and not all specific modes within a general area are discussed. References for more information are provided at the end of each segment.

6.2.1 Plastic Deformation

Plastic deformation is the permanent distortion that occurs when a material is subjected to tensile, compressive, bending, or torsion stresses that exceed its yield strength and cause it to elongate, compress, buckle, bend, or twist. Figure 6.2 shows schematics of these deformation mechanisms. If the applied stress is high enough, the deformation will continue until cracks start to form. These cracks then propagate until the material fractures completely.

A material does not have to fracture completely for a product element to fail. For example, a small amount of deformation can be enough to put a product element out of proper alignment with other product elements, causing the product to stop functioning. An example of this is a bent motor shaft will be unable to properly transmit power to the load.

More information about plastic deformation of materials can be found in Dieter (1986), Farag (1997), and Hertzberg (1995).

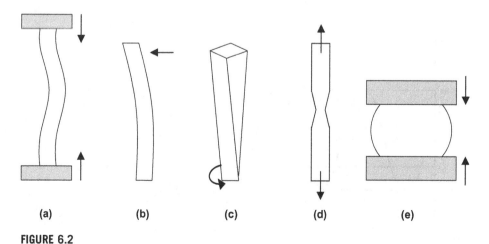

(a)   (b)   (c)   (d)   (e)

FIGURE 6.2

Schematics of (a) Buckling. (b) Bending. (c) Twisting. (d) Elongation. (e) Compression.

6.2.2 **Brittle Fracture**

Brittle fracture occurs in materials that can withstand little or no plastic deformation before fracturing. Brittle fracture occurs in a material when the applied stress exceeds its fracture strength. There is minimal or no degradation of the brittle materials prior to fracture. Ceramics, brittle plastics, and certain high-strength steels are examples of materials that fail by brittle fracture.

6.2.3 **Fatigue**

Fatigue occurs when a material is subjected to repeated cycling between upper and lower stress level. Figure 6.3 shows an example of an idealized stress cycle used for laboratory material testing, where σ_m is the mean stress, σ_a is the stress amplitude, and $\Delta\sigma$ is the stress range. The upper and lower stresses can be both tension (positive), both compression (negative), or one tension and one compression. In actual product use situations, the stress cycles will not be as smooth and repeatable as shown the figure. However, the degradation of the materials will be similar.

Exposure to the cyclic stresses leads to progressive structural damage within a material, which occurs at stresses that are lower than the yield strength of the material. The mechanism of fatigue damage consists of a few steps. First, microcracks form in the material as it cycles between the upper and lower stress levels. As the cycling continues, the microcracks propagate through the material and coalesce, forming one or more larger cracks. As the cycling continues, the larger cracks continue to propagate through the material. Finally, the material fractures completely after the cracks have propagated through so much of the material that the cross-section of intact material is too small to bear the applied load.

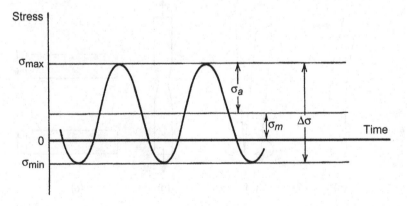

FIGURE 6.3

Fatigue stress cycles used for laboratory testing of materials.

FIGURE 6.4

Fatigue fracture surface from a metal bolt. (*Source: ASM Handbook*, Volume 11, 2002. Reprinted with the permission of ASM International.)

The crack propagation that occurs with each stress cycle appears as striations across the surface of a fractured sample. Figure 6.4 shows the fracture surface from a metal bolt that experienced fatigue damage. The images in Figure 6.5 show the striations at the fracture surface of an aluminum alloy that was fatigue tested in a laboratory and on the fracture surface of a polycarbonate plumbing fixture.

Fatigue cracks typically initiate near or at stress concentrators that lie on or just below the surface of a material. Stress concentrators cause the local stress near the defect to be greater than the average stress on the material. They include sharp changes in cross-section, surface scratches, surface pits, inclusions, embrittled grain boundaries, and preexisting microcracks that form during product element formation. However, even in a flaw-free metal with a highly polished surface and no stress concentrators, a fatigue crack may form if the stresses are high enough.

Sources of cyclic stresses include repeated mechanical loading and unloading, vibration, repeated heating and cooling (i.e., thermal cycling). Cyclic stresses due to thermal cycling occur when materials in contact expand and contract as they are heated and cooled. If the materials in contact have different coefficients of thermal expansion, then during the heating and cooling cycles, the materials

(a)

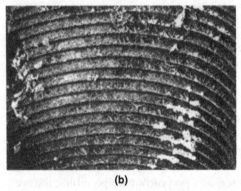

(b)

FIGURE 6.5

Fatigue striations on the fracture surface of (a) an aluminum alloy and (b) a polycarbonate plumbing fixture. (*Source:* (a) *ASM Handbook,* Volume 11, 2002; (b) *Characterization and Failure Analysis of Plastics,* 2003. Reprinted with the permission of ASM International.)

expand and contract by different amounts. The difference in the amount of expansion and contraction as the materials are heated and cooled puts stress on the joint. This is an important consideration for joints and coatings.

Rolling contact fatigue is caused by the repeated mechanical stresses that occur as one body rolls on another body, such as with ball bearings, ball screws, and gear teeth. The damage starts as cracks below the surface and eventually results in surface pits and wear particle debris on the surface of the damaged component. In metals, nonmetallic hard particles and voids lying below a component's surface are stress risers that act as initiation sites for rolling contact fatigue. This fatigue can still occur in the absence of these defects, but more cycles are required.

Mechanical or thermal cycling can cause failure of an interface, such as for a coating on a material or in a composite material between the matrix material and the embedded material in a composite. If the adhesion between the two materials in contact is weaker than the cohesive strength of either material, then repeated mechanical or thermal cycling can lead to a loss of adhesion.

Most ceramics are sensitive to thermal shock and thermal fatigue. High thermal stresses are generated throughout a ceramic body because of variations in the temperature during rapid cooling or heating. These variations result in differences in thermal expansion or contraction throughout the material, which causes internal stresses within the material. These stresses cause existing cracks to grow or new cracks to form and then grow. A single severe thermal cycle is referred to as thermal shock. A large number of thermal cycles that produce more moderate stresses cause incremental crack growth, which ultimately causes complete fracture. This is referred to as thermal fatigue.

More information about fatigue of materials is available in *ASM Handbook*, Volume 19 (1996); Dieter (1986); Farag (1997); Hertzberg (1995); and Munz and Fett (1999).

6.2.4 Creep

Creep is a time-dependent deformation process that occurs when a material is exposed to mechanical loads at elevated temperatures. The term *elevated temperature* is a relative one and varies from material to material. In general, creep is an issue for metals and pure ceramics above about half the absolute melting temperature, measured in degrees Kelvin. For polymers, creep is a concern at temperatures above the glass transition temperature.

When creep is active, the material deforms at a rate that depends on the material, the stresses on the material, and the temperature. The stress levels required to cause creep are lower than the room temperature yield stress of a material. The general creep behavior for materials is shown in Figure 6.6. The amount of deformation over time (i.e. strain rate) increases as the temperature or stress increases. Fracture occurs when the material has undergone so much deformation that it can no longer bear the applied load. Creep fracture is also known as stress rupture.

In metals and ceramics, the creep resistance increases as the grain size increases. Also, for ceramic materials with a secondary glass phase at the grain boundaries, the creep temperature depends on the softening point of the secondary phase. The temperature for creep decreases as the softening point decreases.

More information about creep can be found in Dieter (1986), Farag (1997), and Hertzberg (1995).

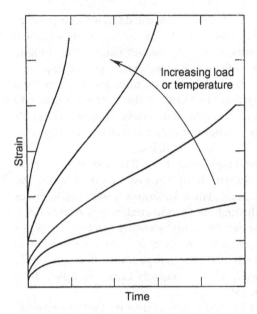

FIGURE 6.6

General creep behavior.

6.2.5 **Wear**

Wear occurs when material from the surface of a product element is removed as a result of the mechanical action of another solid or liquid rubbing against it. Some of the many wear mechanisms are described next and shown schematically in Figure 6.7.

Abrasive wear. Abrasive wear occurs when hard particles or hard surface asperities are forced against and move along a solid surface. The interaction results in removal of material from the surface in contact with the hard particles or asperities. This process is shown in Figure 6.7(a).

Adhesive wear. Adhesive wear is caused by metal-to-metal contact, resultant welding at the contact, and breaking of the welds. This occurs when surface asperities from one component are pressed against the surface asperities of another component, resulting in welding of the asperities. The welding occurs at temperatures below the melting point of the materials in contact and requires that the normal force to the asperities be high. When one or both of the materials in contact start to move, material is pulled from one of the surfaces involved in the weld. Eventually, the separated material dislodges from the surface and acts as a hard particle, causing abrasive wear. This process is shown in Figure 6.7(b).

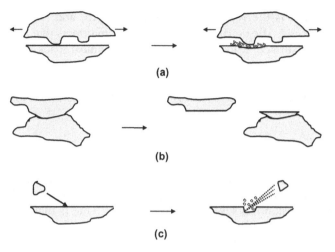

FIGURE 6.7

Schematics of different wear mechanisms. (a) Abrasive wear. (b) Adhesive wear. (c) Solid particle erosion.

Fretting wear. Fretting is caused by small-amplitude oscillatory motion between contacting surfaces. The motion is usually the result of external vibration or micromotion associated with temperature cycling and differences in the coefficient of thermal expansion of the materials in contact. The repeated oscillatory motion leads to the removal of material from the surfaces of the materials in contact. The total range of motion can be on the order of microns. When this process occurs in metals under normal atmospheric conditions, the wear debris can oxidize at the interface between the contacting components. Fretting is a major cause of failure in electrical contacts as the oxide debris accumulates in the contact area, causing poor electrical contact.

Rolling contact wear. This mechanism is associated with the wear damage that occurs as a result of rolling contact fatigue, and it was described in the discussion on fatigue presented earlier.

Impact wear. This mechanism occurs on a solid surface that is exposed to repeated impact by another solid body.

Corrosive wear. This mechanism involves the combined effects of wear and corrosion, which can result in total material losses that are much greater than the additive effects of each process taken alone.

Solid particle erosion. This is the loss of material from repeated impact by small, solid particles. A graphic of the process is shown in Figure 6.7(c). Applications where this is a concern include steam and jet engines and pipelines and valves carrying particulate matter.

Cavitation erosion. Cavitation is the repeated nucleation, growth, and violent collapse of bubbles in a liquid. The collapsing bubbles cause microjets that are directed toward the solid surface, leading to localized mechanical loads that can be so severe that material is removed from the surface.

More information about wear can be found in *ASM Handbook,* Volume 18 (1992), and Engel (1976).

6.2.6 Corrosion

Corrosion is the deterioration of a material as a result of a chemical or electrochemical interaction with its environment. Metals, ceramics, and polymers are all susceptible to corrosion. Corrosion has several potential effects. One is that the thickness of the corroding material reduces, sometimes to the point where the entire thickness is eaten away. This can lead to mechanical failure if the material is exposed to mechanical loads while the material's load bearing area is reduced. Loss of material can also lead to leaks in pipes, welded joints, seals, and around adhesive joints. Another result of corrosion is the buildup of oxide layers on a metal surface. For electrical contacts, the oxide layer will cause high electrical resistance between the surfaces. Finally, corrosion causes deterioration of the cosmetic appearance of a material.

Metals are affected by electrochemical corrosion, gaseous corrosion, and molten salt and liquid metal corrosion; descriptions follow.

Electrochemical corrosion. This mechanism involves the dissolution of a metal or the conversion of the metal to an oxide. The reaction requires that two or more reactions take place on a metal surface in the presence of an electrolyte. An electrolyte is an ionic liquid such as water, an acid, or a base. One of the reactions must involve metal atoms giving up electrons, and another reaction must involve atoms or ions gaining the electrons. The material losing electrons is called the anode, and the material gaining electrons is called the cathode. The flow of electrical current (electrons) from the anode to the cathode is required for electrochemical corrosion to occur. The electrolyte enables the current flow that carries the electrons from the anode to the cathode. When a metal atom at the anode loses electrons, it becomes a positively charged ion. It then dissolves into the electrolyte or reacts with oxygen to form metal oxide on the metal surface. Figure 6.8 shows a schematic of this process.

The extent of the electrochemical corrosion that takes place depends on the particular metal, its physical shape, the materials with which it is in contact, and the chemistry of the electrolyte (e.g., water, type of acid, type of base, pH). Finally, electrochemical reactions can occur uniformly or nonuniformly over the surface of a metal. There are several types of nonuniform electrochemical corrosion mechanisms, including galvanic corrosion, crevice corrosion, pitting corrosion, and filiform corrosion.

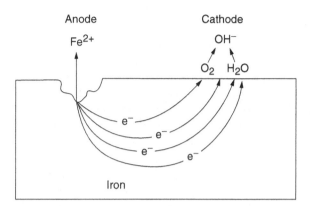

FIGURE 6.8

A schematic of electrochemical corrosion.

Gaseous corrosion. This occurs when a metal is exposed to air or other oxidizing atmospheres at high temperature, resulting in the formation of oxide scale on the metal surface. Gaseous corrosion consists of reactions of a metal with gases such as oxygen, nitrogen, hydrogen sulfide, and sulfur dioxide. Gaseous corrosion is usually a problem for metals only at high temperatures, as the typical reaction rates of industrial metals with common gases are low at room temperature. For metals, the most obvious result of gaseous corrosion is the formation of oxide, nitride, and sulfide scales on a material's surface. The properties of a scale and the development of stresses in the scale determine if it provides a continuous oxidation protection layer or if it cracks and flakes off, allowing the metal underneath to be attacked.

Molten salt and liquid metal corrosion. This results in the dissolution of the metal into the molten salt or liquid metal. In molten salt and liquid metal corrosion, container materials that hold a molten salt or liquid metal are attacked.

For ceramics, oxides and certain silicates are typically stable at high temperature in oxygen or mixed oxygen-nitrogen atmospheres, whereas most of the carbides, nitrides, and borides are not stable in these conditions. The amount of oxygen in the environment affects the type of oxide that forms on these materials. Exposure to hydrogen, ammonia, and carbon monoxide at high temperatures will result in the reduction of oxide ceramics. For example, hydrogen can attack SiO_2 and convert it to SiO. The degree of reduction depends on the stability of the oxide in question.

More information about corrosion can be found in *ASM Handbook,* Volume 13A (2003); *ASM Handbook,* Volume 13B (2005); Fontana (1985); and Uhlig and Revie (2008).

6.2.7 **Chemical Degradation**

Exposure to liquid chemicals can degrade the microscopic structure of polymers. The specific degradation and the extent to which it occurs depend on the material and the conditions under which it is exposed. Dissolution of linear or branched thermoplastic polymers occurs when they are exposed to large enough quantities of certain solvents. In smaller quantities, these solvents are absorbed by the polymer, which can cause plasticization, resulting in softening of the polymer and a corresponding reduction in mechanical strength. The absorption of water can also lead to plasticization in thermoplastics.

Thermoset polymers do not dissolve but swell significantly when exposed to certain chemicals. Swelling also results when rubber products are exposed to chemical solvents. The degree of swelling depends on the material and the solvent. Highly swollen rubbers exhibit a severe loss of strength and stiffness, a high creep rate, and a high rate of stress relaxation, also referred to as compression set.

Exposure to certain chemicals can also cause polymer chains to break into lower molecular weight compounds that no longer have the strength or toughness properties of the original material. This can happen by hydrolysis, when a susceptible polymer is exposed to water or water vapor.

Environmental stress cracking of polymers is a brittle fracture failure mode that results from exposure to mechanical stress in the presence of a chemical that initiates stress relief. This relief results in cracking. The source of the stresses can be internal to the material (residual) or external. Environmental stress cracking occurs at stresses below those required to cause failure in the absence of the chemical. Also, chemicals that induce this failure mode usually have no other apparent effect on the plastic in the absence of a mechanical stress.

More information about the chemical degradation of polymers can be found in Lampman (2003).

6.2.8 **Thermal Degradation**

Exposure to elevated temperatures causes polymer chains in high-molecular-weight polymers to break. The change in molecular weight causes changes in the appearance and mechanical properties of the polymer. Thermal degradation can occur during component fabrication and during a product's use. Many polymers oxidize when heated in oxygen-containing environments. The result is bond breakage, which leads to a reduction in the molecular weight. This is followed by a reduction in the mechanical properties and changes in appearance.

Metals and ceramics exposed to elevated temperatures can undergo phase changes, grain growth, and metallurgical reactions at joints. All of these changes of the microscopic structure cause changes in the properties of the materials.

6.2.9 **Radiation Degradation**

Certain steels exposed to radiation in nuclear reactors can become embrittled. Damage occurs when high-energy neutrons displace metal atoms from their normal lattice positions to form interstitials and vacancies. Also, when neutrons are absorbed by the atoms of an irradiated alloy, reactions occur that produce new metal atoms and gas atoms of hydrogen or helium within the alloy matrix. The effects of damage caused by neutron irradiation include swelling, hardening, and embrittlement. Ceramics used in nuclear reactors also experience swelling.

Ultraviolet radiation from the sun can cause bond breakage along a polymer backbone, resulting in loss of mechanical strength, embrittlement, and changes in appearance.

6.3 CHARACTERIZING THE DEGRADATION AND RELIABILITY OF MATERIALS

It is unrealistic to expect that the materials that constitute a product element will never degrade. Therefore, understanding the physics and extent of material degradation is crucial for designing a product that does not fail prematurely. Characterizing the degradation and its effects is necessary to select materials that enable a product's reliability requirements to be satisfied. Using an inferior material can prove costly in terms of product recalls, delayed product launches, and dissatisfied customers. Conversely, using a material with properties that go well beyond the reliability requirements adds unnecessary cost.

From the standpoint of reliability, some important questions to consider when selecting the materials to use in a product element are as follows:

1. In what manner will the materials degrade when exposed to the use conditions?
2. How fast will the degradation proceed?
3. How will the degradation affect the product element performance?

Information that can answer some of these questions may be available in published technical references. However, much of the published information about the degradation of materials is for specific use conditions and product element physical construction. Since the degradation behavior of a material is influenced by the physical construction of the product element and the use conditions, it is often necessary to perform tests to evaluate the degradation of materials under conditions similar to those seen during product use.

Two approaches are used to evaluate the reliability of the materials used in a product:

1. Product verification testing
2. Material reliability testing

Both types of tests involve exposing test samples to conditions that simulate those that the product will be exposed to during customer use. The performance, material features, or material properties of the test samples are evaluated before and after the exposure and often at intervals during the exposure.

6.3.1 Product Verification Testing

Product verification testing involves exposing samples of a product to simulated product use conditions and then evaluating their performance as a function of the amount of exposure to the test conditions. The purpose is to verify that the product can meet its performance requirements after exposure to the use conditions. Demonstrating that a product can complete the tests and still function as required is a prerequisite that many products must satisfy before being offered for sale.

An example of a product verification test is connecting motors to a load and then turning the motors on and off over a period of time (days, weeks, or months) while exposing them to the maximum temperature and humidity allowed by the motor's product specification. The test samples are placed in a test chamber in which the temperature and humidity can be controlled. The motors' ability to turn the load is monitored during the test.

Product verification tests are performed by Type I and Type II companies at various points in the product development process. One set of tests is performed early in the development cycle to determine the feasibility of the design and to uncover any design flaws that need to be addressed as the development continues. A second set of tests is performed during the detail design phase for the purpose of confirming that any issues that arose during development have been properly addressed and that there are no other major design flaws.

The samples used for the first and second product verification tests are often made using at least a few components and subassemblies produced using manufacturing methods that do not truly represent the methods that will be used when the product goes into production. This means that the materials are probably not completely representative of the materials as they will be when using the intended production processes.

A third set of tests is performed with samples made using the actual manufacturing processes that will be used when the product goes into production. This set of tests is used to verify the reliability of the design and establish that the production manufacturing processes are capable of producing a product that performs as required.

Product verification tests are also performed by Type I and Type II companies when a product already in production goes through redesign or process modifications. The modifications may be necessary for reasons such as cost reduction, performance improvement, or reliability improvement. The purpose of the testing is to verify that the proposed changes have no adverse impact on the performance and reliability of the product.

The data obtained from product verification tests are either pass or fail. Also, information is obtained about changes in a product's performance as a function of the amount of exposure to the test conditions. Test samples pass if the samples are able to satisfy the product performance requirements after test exposure. The test samples fail if they are not able to satisfy the product performance requirements after test exposure.

If the test samples pass, then the design team infers that the features and properties of the materials are at least minimally sufficient to satisfy the product's reliability requirements. If the test samples fail, then the design team must determine the cause of the failure and correct it. If a product fails at a later stage of verification testing, the design team must build new test samples after addressing the root cause of the failure and then test the new samples. It is a design team's goal to pass later phase product verification tests the first time they are performed, especially the final set of tests.

6.3.2 **Material Reliability Testing**

Material reliability testing involves exposing test samples to conditions that simulate those to which the materials will be exposed in a product. The test samples can be individual materials or combinations of materials. Changes in the features, properties, and performance of the materials are monitored during and after the exposure. The test data are used to help the design team understand the relationship between the test sample's physical construction, the test conditions, and changes in the features, properties, and performance of the test samples. Furthermore, the information enables the development of mathematical models that describe the physics of the degradation. These models allow engineers to estimate the reliability of materials, which enables faster identification of suitable materials and physical construction for a product element. If a model is accurate enough, it may be possible to omit material reliability tests for a product element. Nelson (2004) discusses some different models.

An example of a material reliability test is exposing potential cordless telephone plastic housing materials to chemicals or ultraviolet radiation and evaluating the material's strength and color as a function of exposure time and level of exposure. The material used to make a motor shaft can be subjected to laboratory fatigue testing to simulate the mechanical stress cycling that occurs during motor on/off cycles and motor reversals. The fatigue testing can be conducted using various torsion stresses to determine the cyclic loads that a particular material with a particular diameter can withstand.

Examples of the type of data obtained from material reliability testing are shown in Figure 6.9. Figure 6.9(a) shows a graph of the value of a material's feature, property, or performance attribute versus the amount of exposure. Levels 1, 2, and 3 refer to different types of samples or different test conditions. Different types of samples can be the same material with different sample physical construction, different types of materials with the same physical construction, or both. Different test conditions can be different levels of exposure severity for the same type of exposure or different types of exposure. For the case of different test conditions, the conditions in Figure 6.9(a) increase in severity from level 1 to level 3. For the case of different types of test samples and one test condition, the different tests' samples are level 1, level 2, and level 3.

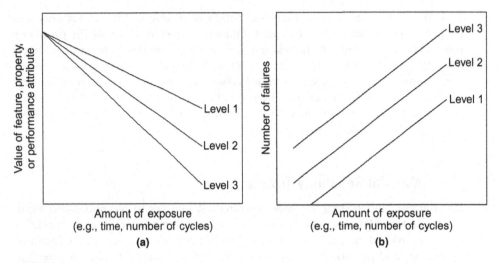

FIGURE 6.9

Examples of data obtained from material reliability tests. (a) Effect of amount of exposure and test severity on an attribute of interest. (b) Number of failed samples as a function of exposure time and test severity.

If there are failure criteria for the materials being evaluated, then the number of failures as a function of exposure can be monitored. Figure 6.9(b) shows a graph of the number of sample failures versus the amount of exposure. Gathering failure data requires that failure criteria be established. The failure criteria can be (1) a change in a property to a value above or below a certain value, (2) a change in the performance to a value above or below a certain value, or (3) a loss of performance. For example, the electrical resistance of a heating element metal can be evaluated as a function of the number of on/off cycles, the maximum operating temperature, and the composition of the test environment (e.g., air or nitrogen). For this example, a failure can be defined as occurring when the resistance exceeds a predefined maximum value, above which a heater element will not perform as required.

The form of a test sample depends on the type of product element, how the materials are used within the intended product element, and the type of information desired. For materials that make up the base material of a component, the test samples may be a simple shape (such as a round or rectangular bar) or a shape similar to the actual component in which the material will be used. Samples of a different component manufactured using the same materials and processes as the one under consideration can also be used. Test samples of coating materials are made by applying them over sheets, plates, or bars of the same material that will be the base material for a component. Joint filler materials can be tested by joining samples of the materials to be joined with the joint filler materials of interest. In-process materials of interest can be tested by applying them over samples of the

Table 6.1 Possible Materials Reliability Tests for the Product Elements Discussed in Chapter 2

Product Element	Exposure Condition	Feature or Property Measured
Wiper blade insert	Ultraviolet light or chemicals	Hardness, flexibility
Motor shaft	Fatigue testing	Fatigue strength
Oven heater element	High temperature and air	Oxidation rate, electrical resistance
Phone solder joints	Thermal cycling	Joint strength, microstructure
Skillet pan	Abrasion with a rough material	Coating wear rate

materials or components that will be involved in the in-process structure. Finally, sometimes the evaluation of the features, properties, or performance of a material requires that the test samples have a specific shape to perform the evaluation. For example, evaluation of the tensile strength of a plastic after exposure to chemicals requires the test samples to have the proper size and shape for tensile testing.

For the products elements discussed in Chapter 2, possible material reliability tests are listed in Table 6.1. As an example of a possible test, consider the impact strength of the plastic material used for a cordless telephone housing. Potential materials could be soaked in different household liquids and chemicals at a specified temperature (e.g., 25°C) for a specified period of time (e.g., 24 hours). Afterward, the impact strength of the samples can be measured and compared to the impact strength of samples that were not exposed to the liquids and chemicals. Similar testing can be done to evaluate the effects of ultraviolet radiation exposure on the plastic's impact strength if the telephone is expected to be used outdoors. For either test, if the impact strength requirement of the phone housing is known, then the materials that meet this requirement after exposure can be identified.

It may not be necessary to perform reliability tests for all of the materials that will be used in a product. Design teams should focus on critical product elements. Failure mode and effects analysis (FMEA) can be used to identify the product elements that pose the greatest risk to the product's performance and reliability and that deserve the most attention. FMEA is explained in Chapter 7.

Tests like those described here have been performed for a wide variety of materials and product applications. Information about these studies and their data is available in technical journals.

6.4 ACCELERATED STRESS TESTING

It can take a long time to obtain the results of exposing test samples to simulated use conditions. To reduce the time required to collect data, product verification and material reliability testing is often performed using test conditions that hasten the degradation of the test samples. This is referred to as accelerated life testing.

Two techniques can be used to accelerate the degradation of test samples: (1) running the samples at a higher use rate and (2) using higher than normal levels for the exposure conditions. Both techniques are discussed next. Other methods of accelerating tests are discussed in Nelson (2004).

Run the samples at a higher use rate. This technique involves increasing the amount of exposure compared to normal product use. One way to achieve this is to expose the samples to the test conditions for a longer period of time compared to normal use. A schematic of this is shown in Figure 6.10. "ON" refers to being exposed to the test conditions, and "OFF" refers to not being exposed to the test conditions. An example of this type of acceleration is exposing a wiper blade insert material to constant UV radiation instead of exposing it only 6 hours per day.

Another way to increase the use rate is to increase the usage cycle rate compared to that used in the product. The usage cycle rate is the rate at which a test sample is cycled on and off. It is increased by decreasing the time between exposure, as shown in Figure 6.11, or decreasing both the ON and

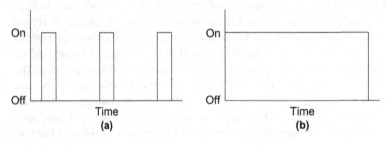

FIGURE 6.10

Increase usage through longer exposure. (a) Normal exposure. (b) Accelerated exposure.

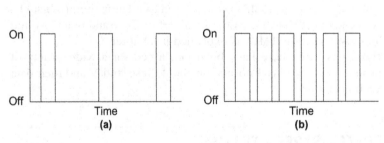

FIGURE 6.11

Increase usage through faster cycling between extremes. (a) Normal use. (b) Accelerated use.

OFF times during testing. A product that normally operates 10 times per day, like a light switch or garage door opener, can be operated 10 times or more per hour to accelerate the testing. The ON and OFF values can also correspond to high and low levels of the exposure conditions. An example of this type of exposure is cycling a heater element wire between 25°C and 1500°C every hour instead of the once per day, or less, which is the case under normal use. The test observes the effects of thermal fatigue on the material's properties and performance.

Use higher than normal levels of exposure. This technique involves using test conditions that exceed those seen during normal operation, such as using a higher temperature, voltage, current, or mechanical load. A motor can be used to move a load two times the normal conditions. Fatigue tests on the motor shaft material can be done using torsion loads that are multiples of the stresses observed during normal use.

Selecting the proper test conditions is an important decision when developing accelerated stress tests. It requires some understanding of the physics of the effects of the test conditions on the materials being tested. The test conditions must not be accelerated in a manner that causes degradation or failure modes that do not correspond to those that would occur under normal use conditions. If this happens, then the test results will not represent the actual product application. The danger is making decisions based on inapplicable test data.

Many accelerated tests are performed that involve the acceleration of more than one test condition. For example, high-temperature corrosion testing of a metal or ceramic can involve using higher than normal temperatures and a corrosion environment that is more aggressive than expected during product use.

Technical journals contain a great deal of published data on accelerated stress testing of various materials. Nelson (2004) contains a bibliography that lists where some of this information can be found.

6.5 ADVANTAGES AND DISADVANTAGES OF PRODUCT VERIFICATION AND MATERIALS RELIABILITY TESTING

Product verification and material reliability testing each have advantages and disadvantages with respect to the following:

- When the tests are performed
- The form of the test samples
- The data collected
- The decisions that can be made based on the test results
- The resources required to perform the tests

6.5.1 **Product Verification: Advantages**

Because the test samples are the complete product, all of the materials are tested at the same time. Also, using the entire product as a test sample enables the design team to evaluate the effects of a product element's physical construction and interactions between materials on material degradation.

6.5.2 **Product Verification: Disadvantages**

From the materials engineering perspective, there are several disadvantages to product verification tests as a method for gaining knowledge about the reliability of the materials. First, little information is obtained about the materials used in a product. If the performance of product verification test samples does not degrade below the minimum product requirements, then the samples are deemed to have passed the test and the materials used in the product are judged to be acceptable. However, if the materials are more reliable than needed, then the product may be more expensive to construct than necessary. If the materials are barely reliable enough, then typical variations in the materials and product use may lead to product failures after the product goes into production.

Second, product verification provides performance and reliability information about the entire product, not individual materials. Only if a product verification test fails is any information about a particular material obtained. Little knowledge is gained about the degradation and failure mechanisms of the materials, and usually the information is not quantitative.

Third, it allows the design team to evaluate few or only one material option for each product element. The cost and effort required to build and test samples for product verification testing typically make it unrealistic to evaluate more than one material option for a particular product element in a product.

Finally, using only product verification puts a design team at risk of having to redesign and retest a product after failing the test. This process is expensive, diverts valuable engineering resources from other products, and adversely affects a design team's ability to meet market timing.

6.5.3 **Materials Reliability Testing: Advantages**

There are several advantages to materials reliability testing. First, materials reliability testing does not require building the entire product. Thus, degradation and reliability information can be gathered in parallel with the other elements of product development. The data can be ready when the design team is ready to select materials. The data will improve the probability of selecting materials that enable the product to pass the final verification tests the first time.

Second, the design team can evaluate several materials options or suppliers of materials at one time. This enables the design team to consider a range of materials options, improving the likelihood of finding a material that meets all of the design requirements for a particular product element.

Third, the test data are quantitative, which allows material features, properties, reliability, and cost for the materials to be compared. Also, physical models can be developed for predicting the degradation behavior of materials exposed to different use conditions. This information can be used to make material selection decisions for future product design efforts or for modifications of the current product, to establish tiers of materials to use for products with different reliability requirements, to help select suppliers based on the features and properties of their materials, and to write specifications for materials.

6.5.4 Materials Reliability Testing: Disadvantages

The main disadvantage to materials reliability testing is extra resources are required to perform the tests. However, the reduced design times, increased certainty of passing product verification tests, and the greater number of materials options are all benefits that justify the extra investment.

6.6 TESTING PROTOCOLS

Product verification and material reliability tests must be carefully designed and performed to collect reliable data. This involves making test samples, selecting the sample attributes to measure, defining the test conditions, developing test equipment and methods for performing the tests, controlling the test conditions, and measuring the sample characteristics. In addition, for material reliability tests, the samples must be designed.

To obtain reliable product verification and material reliability test data, the tests must be carefully designed and executed to collect reliable data. This requires the design team to take the following steps:

1. Identify the material features, material properties, and performance attributes to monitor.
2. Identify, create, and control the test exposure conditions.
3. Identify the test methods for evaluating the attributes of interest.
4. Design the test samples.
5. Make test samples using materials, components, and subassemblies that accurately represent those to be used in the product.
6. Verify composition, microstructure, and properties of materials used for test samples.

Each of these is discussed next. The published technical literature contains many detailed examples of materials reliability tests. This information is valuable for helping design tests.

6.6.1 Identify the Properties, Features, and Performance Attributes to Monitor

The team must identify which product element, subassembly, or assembly attributes are critical to the performance of a product. Key product performance attributes are based on the requirements in the product specification. Key material properties and features are based on the requirements of product elements. Some of the key attributes for the product elements, discussed in Chapter 2, are listed in Table 6.2.

Table 6.2 Key Attributes for the Product Elements

Product Element	Key Attribute
Wiper blade insert	Hardness and flexibility
Motor shaft	Fatigue strength
Oven heater element	Oxidation rate, electrical resistance
Phone solder joints	Joint strength
Skillet pan	Coating wear resistance

6.6.2 Identify, Create, and Control the Exposure Conditions

The test conditions must simulate the exposure conditions defined by the reliability requirements established by the design team for a product element, subassembly, or assembly. The definition of these requirements is discussed in Chapter 2. It may be necessary to identify individual tests for each reliability requirement or it may be possible to evaluate more than one type of exposure within a single test. If the design team wishes to accelerate the exposure conditions, then it must identify appropriate accelerated conditions, as discussed in Section 6.4.

In addition to the use conditions, the interactions and compatibility between materials within an assembly or subassembly must also be considered when determining the test conditions. Electrical components can be sensitive to sulfur outgassing from foams and gaskets. Motion of one component against another can result in detrimental wear. Differences in the thermal coefficient of expansion can result in the fatigue of joints or the delamination of adhesives.

Test equipment and procedures must expose the samples to the intended conditions. This requires proper control of the test conditions and instrumentation for monitoring and recording the exposure conditions. Test samples that are underexposed or overexposed might lead to false negative or false positive results, respectively.

6.6.3 **Identify the Test Methods for Evaluating the Attributes of Interest**

For a product, the methods for evaluating performance should be defined in the product specification. For materials reliability testing, it may be possible to evaluate the attributes of interest using test standards developed by organizations like ASTM International and the International Organization for Standardization (ISO). These standards provide guidance for the evaluation of the features, properties, and performance of many materials. However, it may be necessary to develop a new test method if no standard test methods exist. It is important to develop tests that are repeatable, require good control of sample preparation and evaluation conditions, and use reliable measurement equipment.

6.6.4 **Design the Test Samples**

Designing the test samples involves selecting their physical construction. For product verification, the physical construction is the product itself. For materials reliability testing, the physical construction of the test samples depends on the type of testing and evaluations. It can either simulate the product element or have a size and shape required for evaluating a particular attribute.

A product element can be simulated by making test samples that are similar in appearance to the product element but are constructed in a manner that allows for evaluation. Consider the evaluation of different nonstick coating materials for use on the pan of a skillet. Test samples to evaluate the adhesion and wear of the coatings can be made by applying the coatings to the surface of sheets of the pan metal. Furthermore, the tests can include an evaluation of different surface preparations of the metal prior to coating. Another example is the evaluation of the adhesion of silicone adhesives to the plastic material used for a cordless telephone housing. Test samples can be made by applying the different adhesives to strips of the plastic, as shown in Figure 6.12, and curing the adhesives as required. Then, the samples are tested for shear strength or peel strength by pulling or peeling

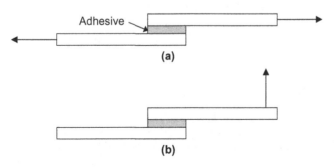

FIGURE 6.12

Adhesion test schematic. (a) Shear test. (b) Peel test.

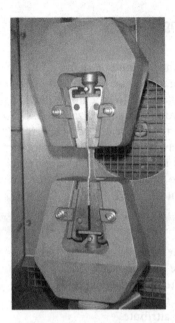

FIGURE 6.13

Tensile test. The test sample is in the grips of the tester.

the strips, as shown in Figure 6.12(a) and (b), respectively. A calibrated force gauge should be used to measure the amount of force required to separate each sample.

Some test samples must have a specific size or shape in order to evaluate a particular attribute. For example, evaluation of the changes in tensile properties of a material may require making samples in the shape of a standard test bar so that its tensile properties can be evaluated using a tensile tester. An example of this is shown in Figure 6.13. The test sample has a specific shape that constrains its deformation to a certain region of the test sample as the grips pull on the sample. This shape may be required if a design team wants to evaluate the effects of conditions such as high temperature or a corrosive environment on the tensile properties of a material.

6.6.5 Make Test Samples

For product verification tests, the materials that make up the test samples must be similar to the materials that the design team intends to use once the product goes into production. This applies to the product elements formed by the company making the product as well as to the components and subassemblies purchased from suppliers. Using materials significantly different from those that are used in a product will result in unreliable data. For example, an epoxy adhesive from one supplier may have worse adhesion than the production-intent epoxy manufactured by a different supplier. Or the fatigue properties of a machined component prototype may differ from those for a die cast component that will be used in the product.

For materials reliability tests, the test sample materials must also be representative of the materials that will be used in the product. This holds for all of the options of materials being evaluated for each product element. However, it is not always necessary to use materials in the exact shape and size as the intended product element. For example, if a design team is interested in evaluating the reliability of weld or adhesive bonds to a custom-designed component from a particular supplier, it is acceptable to use samples from the supplier that are already in production and were made using the same materials and processes as will be used for the component of interest. This is acceptable for components manufactured by various methods (e.g., metal casting, injection molding, and ceramic compaction and sintering). This approach enables materials reliability testing to go forward at the beginning of the design, instead of waiting for prototypes and production tooling.

Off-the-shelf materials, components, and subassemblies should be obtained and used as follows:

- For materials used to form product elements, obtain production samples from the suppliers. When making test samples, these materials must be applied and processed in a manner similar to that intended for production.

- Off-the-shelf components and subassemblies should be the same as those to be used in the product. The concern is that the portions of components and subassemblies that serve as secondary input materials must be representative of what will be used when the product goes into production.

6.6.6 Verify Composition, Microstructure, and Properties of Materials to Be Used for Test Samples

Make sure that the correct materials are tested and that they represent what will be used in production. This means evaluating the composition, microstructure, and significant properties to verify that the materials are as they should be. This may sound obvious, but people often substitute one material for another, one supplier for another, or one process for another without realizing that the substitutes may not have the same features and properties as those that will be used in their product.

6.7 TESTING PROBLEMS

Poor test design and execution result in one of three problems, each of which leads to unreliable data:

1. *Inconsistent test results.* Some samples perform well, and others do not. The properties and performance change appreciably for some samples and not at all for the other samples. When tests are repeated, different results are obtained. The principal consequence of inconsistent test results is the need to repeat the tests, which adds costs and delays to a project.

2. *False negatives.* These are test results that incorrectly indicate that the test samples meet the design's specific requirements. This occurs when the test conditions are not aggressive enough or the test samples are more reliable than the samples that were made using production-intent materials and manufacturing processes.

3. *False positives.* These are test results that incorrectly indicate that the test samples do not meet the design's specific requirements. This occurs when the test conditions are too aggressive or when the test samples are less reliable than the samples that were made using production-intent materials and manufacturing processes.

Inconsistent results occur because of (1) inconsistent test conditions from sample to sample within a test or between tests and (2) inconsistent material features and properties from sample to sample. Inconsistent test conditions can occur because of variations in the test equipment from sample test position to sample test position or poor control over the test conditions. Inconsistent material features and properties result from poor control of the materials used to make test samples or poor control of the processes used to make the test samples.

False negatives and positives occur for a few different reasons. First, design teams might select inappropriate test conditions because the team members do not understand the relationship between the test conditions and the product use conditions. Second, problems occur when the materials and processes used to make the test samples are not representative of the materials and processes to be used in production. This sometimes occurs because the materials and processes of interest are not available and substitutions are used for making test samples. Another reason is poor control or excessive control of the materials and processes used to make the test samples. Poor control results in materials with inferior properties. Excessive control results in materials with properties that are superior to the output from normal production processes.

Test results that incorrectly indicate that the test samples meet the design requirements may cause a design team to select a material that does not have the required reliability. This leads to product failures during customer use, which can lead to customer complaints, reduced sales, and product recalls. Test results that erroneously indicate that a material is unreliable will lead a design team to reject the material, thereby incorrectly eliminating a viable alternative. If the material would have been the best-performing or lowest-cost option, then profits and a competitive advantage are lost.

REFERENCES

ASM Handbook, Volume 11: Failure Analysis and Prevention, ASM International, 2002.

ASM Handbook, Volume 13A: Corrosion: Fundamentals, Testing, and Protection, ASM International, 2003.

ASM Handbook, Volume 13B: Corrosion: Materials, ASM International, 2005.

ASM Handbook, Volume 18: Friction, Lubrication, and Wear Technology, ASM International, 1992.

ASM Handbook, Volume 19: Fatigue and Fracture, ASM International, 1996.

Lampman, S., *Characterization and Failure Analysis of Plastics*, ASM International, 2003.

Dieter, G.E., *Mechanical Metallurgy*, 3rd edition, McGraw-Hill, 1986.

Engel, P.A., *Impact Wear of Materials*, Elsevier, 1976.

Farag, Mahmoud M., *Materials Selection for Engineering Design*, Prentice Hall, 1997.

Fontana, M.G., *Corrosion Engineering*, McGraw-Hill, 1985.

Hertzberg, R.W., *Deformation and Fracture Mechanics of Engineering Materials*, 4th edition, John Wiley and Sons, 1995.

Munz, D., and T. Fett, *Ceramics: Mechanical Properties, Failure Behavior, Materials Selection*, Springer-Verlag, 1999.

Nelson, W.B., *Accelerated Testing: Statistical Models, Test Plans, and Data Analysis*, John Wiley and Sons, 2004.

Uhlig, H.H., and R.W. Revie, *Corrosion and Corrosion Control*, 4th edition, John Wiley and Sons, 2008.

Product Planning and Control Documents

7

7.1 INTRODUCTION

Chapters 7 through 11 discuss the application of the materials engineering perspective to the various phases of product development and manufacturing. The discussions cover the materials engineering considerations and the information required to make well-informed decisions, to identify the risks of failure, and to learn strategies for mitigating the risks.

As mentioned in Chapter 1, although several perspectives must be considered during product development and manufacturing, only materials engineering is considered here. Furthermore, the many elements of product development and manufacturing which do not require the materials engineering perspective will not be discussed.

This chapter examines the materials engineering perspective as it applies to the planning phase of the product development process. During the planning phase, a market opportunity is defined for a new product. This phase is applicable to Type I companies. It is not applicable to Type II and Type III companies because they do not have to define the market opportunity for their product. Instead, Type II and Type III companies rely on a set of design requirements or set of designs, respectively, which their customer provides for the item that they will design and manufacture.

As the development process progresses, a design team writes control documents, which identify all of the requirements that must be satisfied for the product and its subassemblies, components, materials, manufacturing processes, and test methods. These documents are the output of product development, with design teams filling in the requirements as the design process proceeds. This chapter includes a discussion of control documentation, because at the end of the planning phase, if a design team decides to continue with a new product, then the design team must be aware of the requirements information that must be obtained to complete the control documents.

Planning Goals

General Goals
  ■ Define the business opportunity, and determine its feasibility.
 	
Materials Engineering Goals
  ■ Determine the feasibility and risks for finding materials and
    manufacturing processes that will enable a successful product.

7.2 PRODUCT PLANNING

During the planning phase for a new product, a company must perform the following steps:

1. Define a market opportunity (and write it up in a market research document).
2. Determine if the opportunity suits the company.
3. Determine if the company has the resources to develop a product which meets all performance, cost, and schedule requirements.
4. Determine if all the risks associated with developing the product are acceptable.

These steps are listed in the flowchart shown in Figure 7.1.

If the response to any of items 2 through 4 is no, then the marketing and design teams must decide whether to redefine the market opportunity or cancel the project. If the response to all of these considerations is yes, then the teams move on to the next phase. From the business perspective, the purpose of going through the planning phase is to ensure that the company is capable of developing a product that successfully satisfies the market opportunity.

From the materials engineering perspective, the ability to answer the suitability, resource, and risk concerns requires the following information:

- Potential materials and manufacturing processes that are likely to meet the product concept requirements
- An understanding of the product development risks associated with the materials and manufacturing processes.
- Strategies that will effectively mitigate the risks

The remainder of this section discusses the application of the materials engineering perspective for each step shown in Figure 7.1.

7.2.1 Define the Market Opportunity

The company must plant a seed for conceptualizing the design. This is most often a marketing activity and involves developing a marketing requirements document

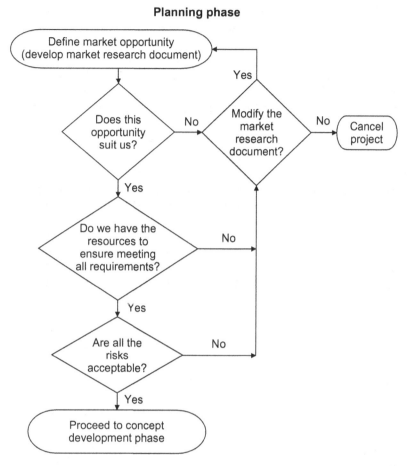

FIGURE 7.1

Flowchart for the planning phase.

(MRD). Frequently, especially for companies with years of experience in specific market segments, this "seed" is presented on a single sheet of paper. It describes, at a minimum, how big the product should be, its basic shape and functionality, the target customer, how much the customer would be willing to pay for the product, how much it should cost to produce, and when it needs to be released onto the marketplace (i.e., launched). The MRD becomes the framework for decision making that follows throughout the product development life cycle.

The MRD sets expectations for what the new product might look like and how much revenue it might lead to. However, it also sets expectations for how the product development, manufacturing, sales, and distribution of the new product will be the same or different from what the company has done in the past. In this sense, the MRD sets expectations for the utilization of people, facilities, and financial resources. As information is acquired during the product development process, it must be compared to the expectations established by the MRD so as to minimize surprises. Any information that challenges the viability of the product design or that reduces the expected return on investment must be evaluated to determine if the project should be canceled or if expectations should be revised accordingly. Finally, the MRD contains a basic analysis of competitors' products and why they do not satisfy the needs of the targeted market.

From the materials engineering perspective, it is important to consider the materials and manufacturing processes that can be used to make the product. If the concept is an iteration of past designs, one cannot assume that the materials will be the same as in past products if any of the new product's design requirements differ from those of past products. Some research should be performed to get a basic understanding of whether past materials and processes can be used. For a new product platform or fundamentally new product, the design team must get a general sense of the materials and manufacturing processes that might be used. In either case, the design team must identify the risks of using the materials and processes and identify ways to mitigate the risks. The risks include the product not meeting requirements because the materials do not have the necessary properties or the product is more difficult to manufacture than anticipated.

The preceding analysis can be more easily applied to product concepts for which most of the subassemblies and product elements will be designed by the company writing the MRD. In this case, the company has control over the materials used to make the product. For complex products that will involve many Type II companies that design major subassemblies, the analysis can be applied only to the materials and processes under the control of the company writing the MRD.

7.2.2 Determine Whether the Opportunity Suits the Company

After developing the MRD, the company must determine if the opportunity suits its business structure. That is, does the company have the expertise and capabilities required to make the product to be successful? The company must determine if it has the following traits:

1. The required technical expertise to develop and produce the product
2. An understanding of how to sell to the target market and the resources required to sell to the intended market
3. The manufacturing capability required to manufacture the product at the estimated volume and with the required quality
4. Experience dealing with the supply chain and supply chain issues associated with the product

If any one of these areas is deficient, then the company can decide to augment its capabilities and expertise, modify the MRD, or cancel the project. It can also employ service companies to supply the expertise or capabilities that the company itself lacks. The MRD can be modified so that the market opportunity better suits the company. However, it cannot be modified so much that the concept no longer represents a product that will satisfy the customer's wants and needs. If neither of the first two options is acceptable, then the company should cancel the project because the risks of failure are too high.

From the materials engineering perspective, a design team must determine whether it has the necessary expertise with the materials that might be used in the product and the proposed manufacturing processes. If it does not, then the team must determine the cost and time required to obtain the necessary expertise.

7.2.3 Determine Whether the Company Has the Necessary Resources

Even if a company has all of the right people and manufacturing processes, the product team must determine if there is sufficient staffing, cash flow and savings, and manufacturing capacity to take on another project. If it is determined that there are insufficient resources to successfully develop and manufacture a new product, then the product team can either cancel the project or find a way to acquire the necessary resources.

The materials engineering resources required for product development are related to the following tasks:

Materials selection. This involves reviewing published technical information about materials performance and reliability, analysis of the features and properties of product elements and process input materials, and materials reliability testing.

Making and testing prototypes. This involves test development and evaluation of the materials to verify their material features and properties.

Manufacturing process development. This involves determining the effects of process inputs on process outputs, selection of the process inputs, and developing output analysis test methods.

Supplier evaluation, selection, and development. This involves evaluation of the materials in suppliers' products and providing guidance to suppliers about how to improve their manufacturing process capability and product quality.

The resources required to perform these tasks depend on the following factors:

- The type of development project (i.e., incremental, derivative, new platform, or fundamentally new product)
- The complexity of the product
- The severity of the product's reliability requirements
- The number of manufacturing processes required to make the product
- The complexity of the manufacturing processes
- The team's familiarity and experience with the materials and manufacturing processes

The costs to select materials for a product can be large if the number of unique components, joints, and in-process structures is large or if many of the materials selection criteria are new to a design team.

The type of development project influences the familiarity with the materials and manufacturing processes used to make a product. For incremental or derivative products, typically there will not be many new materials and manufacturing processes to consider, so the resources to identify materials options and conduct process development will not be significant. The required materials engineering resources will be greater for new platforms and fundamentally new products for which many of the materials and processes being considered are unfamiliar.

However, when developing a derivative product, with different design requirements compared to previous products the team cannot assume that all of the materials used in previous versions of the product will be appropriate for the new product. Instead, the materials used in past products should be evaluated based on the new requirements, and if necessary, new options of materials must be identified and evaluated. For example, a design team assumed that the changes in a verification test for corrosion were minor. The product, built using the same materials that passed corrosion tests required for previous products, failed the revised test. The team discovered that new materials needed to be considered. The test failure put the project behind schedule as the design team evaluated the new materials and repeated the verification tests.

It may not be realistic to expect to completely understand the materials engineering resources that will be required. There are still many unknowns, especially for a new platform and fundamentally new product. However, if the product team has determined that the product suits the company's technical capabilities, then it should be possible to make a rough estimate of the costs and staffing required. By doing so, the company will avoid having to deal with unplanned costs when it realizes that materials engineering support is required.

7.2.4 Determine Whether There Are Any Unacceptable Risks

The risks of concern here include anything that can cause the costs to exceed the budget, delay the development schedule, or prevent the product from meeting the customer's wants and needs. Based on the earlier discussions in this section, at this point in the product development process the sources of risks are associated with (1) the suitability of the product to a company's expertise and capabilities and (2) the resources available at the company. More risks will be encountered from other sources as the project moves along.

From the materials engineering perspective, the major risks are associated with finding suitable materials, developing capable manufacturing processes, and selecting capable suppliers. A design team must be able to realistically assess the existing expertise, capabilities, and resources that can be brought to bear on these three items and whether they really can be achieved in the budgeted time frame. If the team has the available expertise, capabilities, and resources, then some degree of uncertainty with respect to the options of materials, processes, and suppliers can be tolerated. If the team ignores these considerations, then problems and struggles should be expected.

As noted earlier, if some of the risks are unacceptable, the product team can decide to either modify the MRD or cancel the project. Modifying the MRD is an option only if the changes do not result in a product that no longer satisfies a market opportunity. Canceling a project can be difficult, but as said earlier, if the company cannot make a profit, then why go through the effort?

Control Document Goals

General Goals
- Document all of the requirements that must be satisfied to generate a product that meets the wants and needs of the customers.

Materials Engineering Goals
- Document all of the requirements that must be satisfied to form product elements that consistently meet their design requirements.

7.3 CONTROL DOCUMENTS

A control document contains the requirements for an item or process. The item can be a product or its subassemblies, product elements, or materials. The process can be a manufacturing process, analysis procedure, packaging procedure, shipping procedure, or storage procedure. A control document prescribes requirements that must be satisfied and methods that must be followed to ensure that each sample of the product satisfies its design requirements.

The control documents are the blueprints for making a product. Without control documents, not even the simplest product can be expected to consistently meet its design requirements. One important goal of a product development process is consistency—when addressing design requirements; when translating requirements into materials and process candidates; when determining performance acceptability; and when producing each product element, subassembly, and assembly.

Specifications help direct a product design and development effort. Once design and manufacturing teams identify what information must be in a specification, they must work toward obtaining and inserting the information. This focuses the teams on obtaining the information needed to define all the design and manufacturing requirements.

Control documents must clearly communicate all of the requirements that must be satisfied for a product to meet its design needs. The thoroughness and accuracy of a control document will have a large impact on the consistency with which a product meets these requirements, as well as the cost to produce it. As the product's complexity increases, the need for thorough and accurate control documents increases.

As mentioned in Chapter 1, from the materials engineering perspective the success of a product depends on the proper selection and control of the materials that make up a product. The information contained in the following documents helps select optimum materials and control their properties:

1. *Product specification.* Lists the requirements that a product must satisfy.
2. *Subassembly specification.* Lists the requirements that a subassembly must satisfy.
3. *Component specification.* Lists the requirements that a component used within a product must satisfy.
4. *Material specification.* Lists the requirements that a primary input material must satisfy.
5. *Manufacturing process specification.* Lists the required process inputs and process steps.
6. *Analysis methods specification.* Lists the required equipment, sample, and steps for analyzing the process output.

The information that should be placed in each of these documents is discussed in detail later in this section.

Not all of the specifications listed here are relevant for all products or for every type of company. For example, a Type II company will not write a product specification, and a Type III company will only write manufacturing process and analysis method specifications.

Because it can be a tedious task, a certain degree of engineering discipline is required to write clear, detailed specifications. The benefits associated with well-written specifications and the problems associated with poorly written or nonexistent specifications are discussed in this section. Information about writing specifications is provided in McRobb (1989) and Fitchett and Haslam (1988).

7.3.1 Goals

The general goal of writing control documentation is to record all of the requirements for generating a product that will meet the wants and needs of customers. From the materials engineering perspective, the goal is to write specifications that communicate how to form product elements that consistently meet their design requirements.

7.3.2 **Setting Limits for Attribute Values**

As discussed in Chapter 2, a design team must select the limits for the values for each attribute controlled by a specification. The limits can be stated as a range, maximum value, or minimum value. An example of a range is the length of a component expressed as the target length plus or minus the maximum allowable deviation from the target (e.g., 6.05 ± 0.10 cm). An example of a maximum value is the maximum electrical resistance allowable for an electrical conductor (e.g., <2 ohms) or the required amount of impurities in a ceramic (e.g., <0.1 weight percent SiO_2). An example of a minimum value is the minimum thermal conductivity for a component used as a heat sink (e.g., >10 W/m·K) or the required minimum hardness for a metal used for a wear application (e.g., >30 Rockwell C).

The limits for an attribute must be based on the product's design requirements and the capabilities of the process that will be used to form a product element. There must be a balance between the design of the product and the ease of manufacturing a product and its components and subassemblies. If any of the requirements are too restrictive, then it may be difficult or costly to make or obtain the material, component, subassembly, or product. This can be the case if the requirements for an attribute has a range of values that is very narrow, has a very low maximum, or has a very high minimum.

7.3.3 **Additional Functions of Control Documents**

In addition to serving as product blueprints, control documents serve a few other functions, as follows.

They provide a means to record the design and manufacturing decisions. Modifications are made to products after they go into production to improve their quality and to reduce their cost. Specifications provide a reference point of what was done to obtain the current product performance, reliability, total cost, and manufacturing yields. This information gives design and process teams a starting point for making changes to a product's design or its manufacturing processes. It also enables the teams to correlate the changes with any improvements or degradation in the product's performance, reliability, total cost, and manufacturing yields.

They are reference documents for training and addressing problems. Manufacturing teams use process specifications as a tool for training and guiding the operators who work on a manufacturing line. Because they are available to the entire engineering organization, control documents make it possible to communicate the requirements to anyone who needs the information. Good communication of all of the relevant requirements depends on writing detailed specifications. This is especially important as people join an organization and need to understand all of the requirements associated with a product. When dealing with supplier and manufacturing quality problems, the information in specifications helps identify any deviation from the requirements that may have occurred.

They are a communication tool and contract when working with suppliers. A specification for a subassembly or component communicates the relevant design requirements to the supplier of the item. Detailed specifications imply the technical capabilities required to design and make the item. Suppliers who are not capable may be less willing to take on the project, which will spare the team from wasting time with incapable suppliers.

As a contract, a specification holds a supplier responsible for producing an item that meets the design requirements. If relevant information is missing from a specification and the supplier does not have the capability or interest to validate it, then the chances that the item will meet its design requirements are reduced. In some cases, an item may meet the design requirements listed in the specification yet the product will fail to perform as required. In this case, the client company must accept responsibility for the poorly written specification and the problem.

7.3.4 Engineering Tools for Determining Control Document Information

A few engineering tools are available for helping design teams identify all of the relevant requirements to include in specifications and determine the values for individual requirements. Two of these tools are quality function deployment and failure mode and effects analysis.

Quality function deployment (QFD) is a systematic group of activities used to help marketing and design teams (1) identify the customer's wants and needs; (2) understand the importance of these wants and needs and the current level of customer satisfaction with products already available; (3) translate the customer's wants and needs into engineering requirements; and (4) set priorities for the functions, performance, reliability, and manufacturing costs for a product based on the company's and customer's priorities.

Because QFD incorporates the voice of the customer in product designs, it increases the likelihood that the final product will satisfy the customer's needs. Also, QFD recognizes that certain types of design and specification trade-offs are most economical if they are made *up front* before the product design has gone too far. Going through the QFD process increases the probability that the trade-offs will result in a product that satisfies customers. Finally, QFD provides a link between the perspectives of customers, design engineers, competitors, and manufacturing. Therefore, QFD can dramatically improve product development efficiency because design and manufacturing problems are resolved early in the design phase. More information about QFD is available in Terninko (1997), Akao (1990), and ReVelle (1998).

Failure mode and effects analysis (FMEA) is a systematic group of activities used to (1) determine how a product or process might fail during use, (2) predict the effects of the failures, and (3) identify the controls that can be put in place to prevent or detect the causes of failures. The causes of failures are any errors or defects in a product that arise from its design and the manufacturing processes used to make it. The failures, potential and actual, are prioritized according to how serious their consequences are, how frequently they occur, and how easily their causes can be detected or prevented.

Failure mode and effects analysis is used during the design stage to avoid future failures. Later it is used for manufacturing process control, both before and during the ongoing operation. Ideally, FMEA begins during the earliest conceptual stages of design and continues throughout the life of the product or service as new failure modes are encountered.

FMEA includes the following four steps:

1. Identify potential failure modes.
2. Identify the effects of the failures.
3. Identify all of the possible causes of each failure mode.
4. Identify prevention or detection controls that can be used to assess the adequacy of a design or process.

As an example, consider the plastic housing for a cordless telephone described in the following sidebar.

Potential failure mode:
   a. The housing cracks after the phone is dropped.

Effects of the failure:
   a. Dirt and other materials can get into the housing and potentially damage the electronics.
   b. There is a poor cosmetic appearance.

Possible causes of the failure:
   a. The plastic selected is too brittle.
   b. The plastic is too thin.
   c. The plastic becomes embrittled during the injection molding process as a result of a suboptimum process or poor process control.

Controls:
   a. Evaluate the impact strength of different plastic, materials and identify plastics that have the required strength.
   b. Evaluate the impact resistance of plastics as a function of thickness. Specify the minimum required thickness for a particular material.
   c. Determine the effects of the injection molding process inputs on the impact resistance of the housing. Also, conduct periodic evaluations of the impact resistance on housing samples after injection molding.

Going through the FMEA process forces design teams to consider the feasibility of their designs. This includes consideration of the values of the significant attributes for each product element, which includes size, shape, performance, materials properties, and material features. The design team must select values that will yield reliable product elements. Also, it must be possible to consistently manufacture the product element with the desired values. Therefore, the design team must understand the capabilities of the manufacturing processes and production lines used to produce the product elements. This is also true when considering the use of off-the-shelf components and subassemblies. Going through the FMEA also forces design teams to better understand the use conditions to which the product and its product elements will be exposed. This information is crucial for selecting materials to use for product elements.

From the materials engineering perspective, an FMEA will indicate the materials attributes that must be investigated, controlled, and monitored to ensure the reliability of the product element in which the material is used. It will also indicate the manufacturing process steps and controls that are required to make a product, subassembly, or component that will consistently meet its design requirements.

More information about FMEA is available in Stamatis (1995) and *Potential Failure Mode and Effects Analysis* (2001).

7.4 CONTROL DOCUMENT INFORMATION

This section discusses the information that should be included in the specifications mentioned in the previous section.

7.4.1 Product Specification

A product specification is a document that contains information about all of the technical requirements that a product must satisfy. This document adds more specificity to the product concept requirements given in the marketing requirements document (MRD) and includes the following information about the product:

1. Intended use
2. Mechanical drawings showing shape and dimension requirements
3. Mass requirements
4. Functions and features
5. Performance requirements
6. Reliability requirements
7. Test methods for evaluating performance and reliability
8. Sustainability requirements
9. Cost requirements
10. Manufacturing constraints

11. Applicable government regulations
12. Applicable industry standards
13. Intellectual property constraints

All of these, except the test methods, were discussed in Chapter 2. The test methods will be discussed at the end of this section. A product specification can also include information about the packaging, shipping, and storage requirements. More information about developing a product specification is in Pugh (1991) and Ulrich and Eppinger (2004).

A product specification is first written after identifying the customer's wants and needs and converting them to engineering requirements. At that point, the design team sets target requirements for the product's performance and reliability. These requirements may evolve as the design develops and the design team gains knowledge about the trade-offs among the design options that influence the product's performance, reliability, and cost. This may cause the team to compromise on certain product characteristics before setting the final requirements. Alternatively, it may reveal that more product functions and greater performance are possible than was previously thought.

A product specification, along with the information in the MRD, provides the framework for all product design, development, and manufacturing decisions. Therefore, it is critical that the information in the specification is thorough and accurate, which is critical in order to select the materials that make up various product elements.

This is because the selection criteria for the materials that constitute the product elements are derived from the product design requirements, as discussed in Chapter 2. If information is missing, then either uninformed material selection decisions will be made or decisions will be delayed while the design team tries to obtain the necessary product requirements information.

Inaccurate product design requirements can be costly. Performance and reliability requirements that exceed the wants and needs of the intended customer lead to the use of materials that are more expensive than necessary for the application. Conversely, performance and reliability requirements that are below the wants and needs of the intended customer will result in an undesirable product that underperforms and has poor reliability, resulting in unhappy customers and low sales.

As an example, if the specification for a windshield wiper blade insert dictates that it meet certain temperature requirements to which the insert will never be exposed, then the materials options for the wiper blade insert will be limited and potentially more expensive compared to an insert with lower temperature requirements. This directly reduces profitability. On the other hand, if the specification identifies temperature requirements that are below those that actually occur, then the insert material that is selected might not withstand the actual use temperatures for the desired use period, diminishing its reliability.

Test Methods for Evaluating Performance and Reliability

Test methods for evaluating each performance requirement must be identified. Industry standard test methods should be used when possible. When it is necessary to develop test methods, a specification should be written for them. The format should be as discussed later for analysis methods specifications.

The test methods used to evaluate a product's reliability should include descriptions of verification tests that the product must pass in order to be sold. The requirements for verification tests should include a complete description of the test conditions. As discussed in Chapter 6, it is important that the tests are physically realistic—that is, that they only cause materials degradation that could occur during actual use. If the tests are too aggressive, then the materials selected may be more expensive than really required for the product's reliability. Furthermore, overly aggressive tests may result in erroneous failures that delay product development. Conversely, if the tests are not aggressive enough, then the product may not have the reliability required and may fail during customer use.

7.4.2 Subassembly Specification

Subassembly specifications are written by Type I and Type II companies for subassemblies to be used within their products. A subassembly specification should contain the following information:

1. Description of the function of the subassembly
2. Scale drawing of the subassembly, including dimensions
3. List of the subassemblies and components within the subassembly
4. Performance requirements
5. Reliability requirements
6. Test methods for evaluating the performance and reliability
7. Applicable industry standards
8. Applicable government regulations
9. Intellectual property requirements
10. Sustainability requirements
11. List of materials used to fill joints and form in-process structures
12. Composition requirements of the materials used to fill joints and form in-process structures
13. Materials' properties requirements for joints and in-process structures
14. Microscopic structure requirements for joints and in-process structures
15. Requirements for the defects within joints and in-process structures
16. Joint and in-process structure performance requirements
17. Joint and in-process structure reliability requirements
18. Test methods for evaluating joint and in-process structure performance and reliability

The specification may also contain information about the requirements for packaging and shipping if the subassembly is to be manufactured at a different location from where the product itself is built.

The first ten items, except for the test methods, were discussed in Chapter 2 and are similar to the items in a product specification. The considerations for the test methods are also similar to those for a product specification, with one addition. The reliability requirements can vary from subassembly to subassembly within a product. In the case of an automobile, for example, subassemblies in the passenger compartment are exposed to temperatures up to 40°C, while some subassemblies in the engine compartment are exposed to temperatures in excess of 125°C as a result of exposure to the heat from the engine.

The last eight sets of requirements are related to the joints and in-process structures and are described next.

List of materials used to fill joints and form in-process structures. The material may be identified by composition, industry standard name, trade name, or brand name. For each material listed, there should be a material specification that discusses the requirements for the material. A reference to the corresponding materials specification should be provided for each material listed. Material specifications are discussed later in this chapter.

Composition requirements of the materials used to fill joints and form in-process structures. The format for describing the material composition depends on the specific material. For metals, the values of each element present in the material are range, maximum, or minimum; whereas, a polymer is described by its chemical name. Materials made by specific suppliers may have trade names that must be included in the specification

Materials' properties requirements for joints and in-process structures. For each materials property requirement, there must be a reference to a test method for evaluating the property. The test method can be an industry standard or written by the design team.

Microscopic structure requirements for joints and in-process structures. Because of their impact on the properties of a material, some aspects of a material's microscopic structure must be specified to ensure that all of the required properties are obtained. As discussed in Chapter 4, it is possible to use a specific material composition but get different microscopic structures depending on the manufacturing processes used to form a product element. Different microscopic structures will result in different material properties and reliability, even for the same composition material.

The details of the microscopic structure should be quantified whenever practical. For features, such as the phases present in a metal or ceramic, it is not always possible to provide a value. Instead, a detailed qualitative description must be provided along with a micrograph for demonstration.

Requirements for the defects within joints and in-process structures. This includes a description of the types of defects that are acceptable along with limits for their size and quantities.

Joint and in-process structure performance requirements. This was discussed in Chapter 2.

Joint and in-process structure reliability requirements. This was discussed in Chapter 2.

Test methods for evaluating joint and in-process structure performance and reliability. The discussions for product and subassembly test methods apply here.

7.4.3 Component Specification

A component specification contains the following information:

1. Scale drawing of the component, including dimensions
2. Description of the function of the component
3. Performance requirements
4. Reliability requirements
5. Composition requirements for all of the materials that make up the component
6. Properties requirements for all of the materials that make up the component
7. Microscopic structure requirements for all of the materials that make up the component
8. Surface condition requirements
9. Requirements for the defects within the materials (including the types of defects, their maximum size, and the maximum number)
10. Test methods for evaluating the component's performance, reliability, material properties, and material features

Except for surface condition requirements, the information for all of these items is similar to the information for joints and in-process structures in a subassembly.

The surface condition requirements must address items such as cleanliness, roughness, and appearance. References to applicable test methods must be provided for each requirement.

The amount of detail for any of these items depends on the particular component and its application. For example, it is probably unnecessary to specify microscopic structure requirements for a metal component with cosmetic and minor structural functions. Design teams must determine the information required to ensure that a component's performance and reliability requirements can be satisfied. The design FMEA process will help with this task.

A component specification should contain enough information so that it is impossible to manufacture a component that meets the specified performance and reliability requirements, but does not work properly in the product. For example, if a die cast metal component will be used in a high reliability adhesive joint, then the specification for the component must include a requirement for the component's surface cleanliness. This requirement will influence the die lubricants that can be used during the die casting process or require that components be cleaned after they are die cast. These lubricants are sprayed into the molds before running the die cast process so that a die cast component can be easily removed from its mold. Lubricant residue can be present on the surface of a die cast component, and can have a negative impact on adhesion to a component.

7.4.4 **Material Specification**

A material specification contains the requirements for the primary input materials used to fabricate a product element. A material specification contains the following information:

1. Approved uses for the material
2. Composition requirements
3. Material properties requirements
4. Microscopic structure requirements
5. Surface condition requirements
6. Allowable defects
7. Analysis methods for evaluating the properties and features of the material
8. Referenced documents

Each of these is described next.

Approved uses for the material. The approved uses are the applications in which a material can be used. For example, this might be for a specific type of product element or for a certain set of performance and reliability requirements. A list of approved uses prevents the material from being used in applications for which it has not been evaluated.

Composition requirements. The format for describing the material composition depends on the specific material. For metals, the values of each element present in the material are range, maximum, or minimum. A polymer is described by its chemical name. Materials made by specific suppliers may have trade names that need to be included in the specification.

Material properties requirements. The list of materials properties should include the properties and the required values for each property. The properties that should be included in a materials specification are those that are important for the performance and reliability of the product element in which the material is used. The specification should also include properties that are important for use of the material as an input material in manufacturing processes.

The value for each property should be expressed as a range, a maximum, or a minimum. The values must be based on the performance, reliability, and manufacturing requirements of the product element in which a specific material is used. Also, the value selected for each property should be within the range of normal variation of the manufacturing process used to produce the material.

Microscopic structure requirements. The details of the microscopic structure should be quantified where practical, with the values being a range, maximum, or minimum. For features, such as the phases present in a metal or ceramic, it is not always possible to provide a value. Instead, a detailed qualitative description must be provided along with a picture for demonstration. The microstructure requirements should correlate with the requirements for the material's properties.

As applied to materials like metal sheet, metal bars, or blocks of ceramic, the microscopic structure consists of grain size, phases present, the size and size distribution of the phases, the relative amount of the different phases, and the distribution of the phases throughout the material. For input materials such as powder, solder paste, or braze compound, the concept of a microscopic structure is broader and includes features such as particle size and particle size distribution.

Surface condition requirements. This includes details about surface cleanliness, roughness, and chemistry. This is important for materials to which a joint is to be formed or coated. In addition, a component may have cosmetic appearance requirements.

Allowable defects. Defects can be internal to the material or on the surface. The specification should indicate the types of defects that are acceptable along with limits for their size and quantities.

Analysis methods for evaluating the material features and properties of the materials. The relevant test method for evaluating the attribute for each composition, property, microscopic structure, surface condition, or defect requirement must be specified.

Referenced documents. This includes all references to other documents such as industry standards or other specifications written by the company. For example, the composition of a material may include a reference to an ASTM International specification that discusses the specific material.

The amount of information in a specification depends on the material and the application in which it is used. The information that is required depends on the material properties that must be controlled to ensure the performance and reliability of the product element in which the material is used and the ease of using the material in a manufacturing or assembly process. Two product elements made using the same material can have different specifications for the material's properties and features, depending on the product elements' requirements and the processes used to form them. For example, a process that involves applying an adhesive paste by hand is more tolerant of variation in the paste viscosity than a process that invovles applying it with an automated piece of equipment. The machine will use a set pressure to apply a certain amount of adhesive; for the manual process, however, the operator can use whatever force is necessary to apply the desired amount of paste. Another example is the specification for sheet metal. Specification of the metal's composition may be sufficient to control the required properties for a material used to form a simple shaped component fabricated using simple metal sheet bending processes and used only for cosmetic purposes. But it may be necessary to specify the composition, microstructure, yield strength, and tensile strength for a material used to form a complicated component fabricated using more complicated bending processes and used for rugged mechanical purposes.

Finally, the values for the properties, composition, microstructure, surface condition, and allowable defects requirements are based on the data from the materials selection and process development processes.

7.4.5 Manufacturing Process Specification

A manufacturing process specification contains information about the requirements for the inputs and outputs and the process steps to follow. Its purpose is to control how an item is to be manufactured and thereby control the material features and properties, performance, and reliability of the process output.

A specification should contain the following information for a particular process:

1. Description of the process output
2. List of process equipment
3. List of primary input materials (including references to the specifications for the materials)
4. Components or subassemblies being joined (including references to the specifications for the components and subassemblies)
5. List of process steps and instructions for each step
6. Equipment process parameter set points
7. Process output variables to evaluate (e.g., product element, subassembly, or assembly performance); they can also be the materials' properties and features for product elements
8. Tests used to evaluate process output (including references to analysis methods specifications)
9. Requirements for the values of the process output variables
10. Responses to output measurements that do not satisfy the specification of the item being manufactured (e.g., a response may be to change the values of one or more process parameter set points)

7.4.6 Analysis Method Specification

An analysis method specification controls how analyses are performed so that the data obtained are reliable. An analysis can measure an item's performance, properties, or materials features. The item can be an input material, component, subassembly, assembly, joint, or in-process structure.

An analysis method specification contains the following information:

1. Description of the analysis
2. Description of the analysis equipment
3. Requirements of the samples that can be analyzed (including the requirements of the physical form of the samples)
4. List of analysis steps and instructions for each one

Sometimes it is possible to use an industry-approved test method instead of developing product-specific test methods. ASTM International and the International Organization for Standardization are examples of organizations that write and publish test methods for material and product analysis.

REFERENCES

Akao, Yiji, ed., *Quality Function Deployment: Integrating Customer Requirements into Product Design*, Productivity Press, 1990.

Fitchett, P., and J. Haslam, *Writing Engineering Specifications*, Spon Press, 1988.

McRobb, M., *Specification Writing and Management*, Marcel Dekker, Inc, 1989.

Potential Failure Mode and Effects Analysis (FMEA), 3rd edition, Automotive Industry Action Group, 2001.

Pugh, S., *Total Design: Integrated Methods for Successful Product Engineering*, Addison-Wesley, 1991.

ReVelle, Jack B., John W. Moran, Charles A. Cox, *The QFD Handbook*, John Wiley and Sons, 1998.

Stamatis, D.H., *Failure Mode and Effect Analysis*, ASQC Quality Press, 1995.

Terninko, John, *Step-by-Step QFD: Customer-Driven Product Design*, 2nd edition, CRC Press, 1997.

Ulrich, K.T., and S. D. Eppinger, *Product Design and Development*, 3rd edition, McGraw-Hill, 2004.

Product Concept Development

8.1 INTRODUCTION

During concept development, the needs of the target market are thoroughly researched, a detailed product specification is developed, and product design concepts are developed. A product concept is a description of the form, function, and features of a product. It is usually expressed as a sketch or rough three-dimensional model. After they are evaluated, one or more product concepts are selected for further development.

Concept development is applicable to Type I and Type II companies. It is not applicable to Type III companies, who manufacture their products based on designs provided by their customers.

Many of the decisions made during concept development have a significant impact on the selection and control of the materials used in a product. This chapter examines the considerations and information required to make informed decisions.

From the business and general engineering perspectives, the concept development phase includes the following steps:

1. Perform a detailed market analysis.
2. Write a detailed product specification.
3. Generate product concepts.
4. Evaluate product concepts.
5. Determine whether any of the concepts are acceptable. If they are not, then the design and marketing teams must decide to modify the concepts, generate new concepts, revise the product specification, or cancel the project because changes to the product specification would result in a product that does not satisfy the customer's wants and needs.

Product concept development

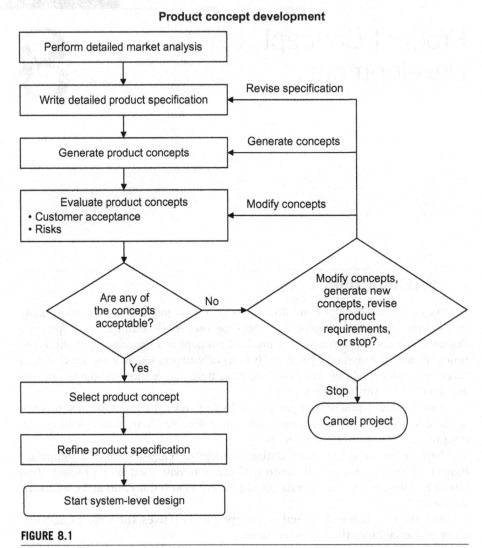

FIGURE 8.1

Product concept development flowchart.

6. Select product concepts.

7. Refine the product specification as required.

Figure 8.1 shows a concept development flowchart that contains these seven steps. Each of these steps will be discussed in this chapter. There are other aspects to this phase that are not shown. However, they are not relevant to this discussion. Discussions about these other aspects of concept development can be found in Pahl and Beitz (1996) and Ulrich and Eppinger (2004).

> **Detailed Market Analysis Goals**
>
> **General Goals**
> - Determine the customer's wants and needs.
> - Understand price, performance, reliability, cost to produce, and the customer's acceptance of competitors' products.
>
> **Materials Engineering Goals**
> - Correlate the properties and features of the materials used in competitors' products to the performance, reliability, and cost of the product elements and the products.
> - Identify options for materials and manufacturing processes that can be used in the product being designed.

8.2 PERFORM DETAILED MARKET ANALYSIS

A detailed market analysis involves obtaining a complete picture of the market opportunity. The purpose is to help a company to better understand the opportunities and risks associated with developing a product. Information from the analysis will help better inform if there is indeed a need in the marketplace for the product proposed in the original marketing requirements document (MRD), if the MRD needs to be modified, or if the project should be canceled. The information also helps marketing and design teams come up with a reliable estimate of the costs and time needed to develop and manufacture a product, determine the price that customers will pay for a proposed product, and identify the risks to success.

Ultimately, a design team needs the information obtained during market analysis to properly assess the likelihood that it can design, develop, and efficiently manufacture a product that cost effectively meets the customer's expectations and earns a reasonable profit. After all, if it is not possible to make a reasonable profit, then why put forth the effort?

A detailed market analysis involves the following steps:

1. Identify the customer's wants and needs.
2. Evaluate competitors' products.
3. Evaluate the customer's interest in the product concept in the MRD.
4. Identify risks and risk mitigation strategies.
5. Determine whether there are unacceptable risks.
6. Revise the MRD as required or cancel the project because changes to the MRD would result in a product that does not satisfy the customer's wants and needs.

The flowchart for the process is shown in Figure 8.2. This can be an iterative process if changes to the MRD are required. These changes can be with respect

Concept development—detailed market analysis

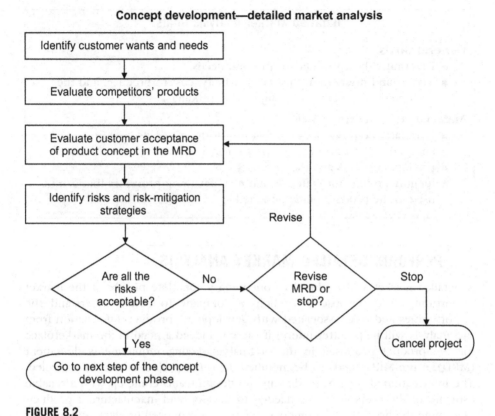

FIGURE 8.2

Detailed market analysis flowchart.

to the sales price, sales volume, or initial product concept. Changes to the product concept require reevaluation of the customer acceptance. Any changes to the MRD require the same analysis as used for the planning phase.

The remainder of this section discusses the steps of the detailed market analysis and the materials engineering considerations.

8.2.1 Identify the Customer Wants and Needs

Identifying the customer's wants and needs is critical to the success of a product. If the set of wants and needs is not complete and accurate, then the organization runs the risk of developing a product that the customer may deem undesirable even though it meets the technical requirements developed by the design team. It is possible to develop a product that works great but does not meet customer expectations because the effort was not taken to thoroughly and accurately understand the customer's wants and needs.

The task of identifying the customer's wants and needs is usually left to an organization's marketing team. The information is obtained using techniques such as competitor analysis, customer surveys, and customer focus groups.

From the materials engineering perspective, identifying the customer's wants and needs is important because it is the first link in the chain that establishes the selection criteria for the materials that make up the product. As was shown in Chapter 2, the customer's wants and needs are converted into technical requirements that will make up the product specification, from which the design requirements for the product elements in a product are derived.

After the wants and needs have been determined, they must be converted to engineering requirements because wants and needs are frequently stated in non-technical terms. For example, stated wants and needs for a cordless telephone might be "good sound quality," "holds lots of numbers in memory," and "looks high-tech." The design requirements shown in Chapter 2 for the automobile, telephone, oven, motor, and skillet are examples of technical requirements that are included in a detailed product specification. As discussed in Chapter 7, quality function deployment is an effective engineering method for converting the customer's wants and needs to engineering requirements.

8.2.2 Analysis of Competitors' Products

Analyses of competitors' products involves examination of the products' features, performance, reliability, architecture, construction, materials, and customer acceptance. This information will help a design team make decisions about the features to include in its product and about the product's performance and reliability. The information obtained from analyzing the construction and materials will help understand the following:

- The materials that enable specific performance and reliability attributes
- The reasons for poor performance and reliability
- The manufacturing processes used to form product elements
- The cost of the product elements

There are two benefits to gaining this knowledge. First, understanding the materials and their impact on the competitors' products helps a design team identify the materials and manufacturing processes that can be used to make their product. In a sense, analyzing competing products is like performing a huge experiment without having to actually make samples. The design and materials used for the different products are the independent variables, and the performance, reliability, and cost are the dependent variables. Furthermore, a design team might determine that different materials and manufacturing processes can be used to create a product that has more favorable performance attributes, reliability, or cost than the competitors' products.

The second benefit is that the knowledge helps the design team better understand the technical expertise and capabilities required to make a product that has features similar to those in the competitors' products, if that is desirable. This helps the design teams better assess the risks associated with continuing development of the new product.

Analysis of competitors' products involves the following steps:

1. Select products to evaluate
2. Evaluate the significant product performance and reliability attributes
3. Characterize the product's architecture and identify the significant product elements and their functional roles
4. Evaluate material features, material properties, performance, and reliability of the significant product elements

Many companies already perform steps 1 to 3. Step 4 addresses the materials engineering goal to help the team understand the relationship between the performance, reliability, and cost of competitors' products and the materials and manufacturing processes used to make them. The four steps are discussed next.

Select Products to Evaluate   This involves selecting products that are competitors to the new product being considered.

Evaluate the Significant Product Performance and Reliability Attributes   As mentioned in Chapters 6 and 7, the tests for evaluating a performance or reliability attribute must be designed so that each test isolates a single attribute, provides quantitative data, can be easily repeated, and gives reliable results.

Characterize the Architecture and Identify the Significant Product Elements and Their Functional Roles   The purpose of this evaluation is to understand how the product functionality was engineered and how the product elements relate to the product functionality. The analysis involves identifying (a) the different high-level subassemblies within the products, (b) the mechanical and electrical interfaces between the subassemblies, (c) the lower-level subassemblies, (d) the product elements within the lower-level subassemblies, and (e) the functionality of the product elements.

The information from the evaluation helps a design team identify the product elements most critical to the functionality of each product and to determine which product elements enable differentiating features. It is important to distinguish the differences among the relative importances of the various product elements within a product. For example, the product elements that are critical for an industrial oven to achieve its advertised thermal performance are the heater element wire, insulation, and door seal. The components that constitute the oven case and door handle are necessary from a use and cosmetic perspective, but they are not critical to the performance of the oven. The depth of understanding about the materials used for the heater element, insulation, and door seal is greater than for the oven case and door handle.

For products that contain off-the-shelf subassemblies, it may not be beneficial to divide the subassemblies into their constituent product elements because an off-the-shelf subassembly is selected primarily based on its overall performance and reliability. The product elements and materials that constitute an off-the shelf subassembly are usually not important to a design team except for two concerns: (a) there may be some restrictions on the materials for environmental compliance or sustainability requirements and (b) information about the components of a subassembly that are secondary input materials (i.e., joined to other subassemblies or components).

Evaluate Material Features, Material Properties, Performance, and Reliability of Significant Product Elements   The design team must identify the relevant material features, material properties, performance attributes, and reliability attributes for the critical product elements. The test methods that will be used for the evaluations must also be identified.

This information helps design teams understand (1) the materials that it should or should not use in its product, (2) the manufacturing processes that should or should not be used to form product elements, and (3) the costs to build the competitors' products. If the performance or reliability for a product is lacking, the materials information may indicate that suboptimum materials were used or that the product elements were not properly formed. Conversely, good performance or reliability will point toward acceptable materials and manufacturing processes. Finally, since the cost to build a product depends on the materials and manufacturing processes, information about these will help estimate the cost to make competitors' products.

Whether it is possible or appropriate to evaluate all of the relevant properties, features, performance, or reliability of a particular product element depends on the form of the product element under consideration. The material properties that can be measured may be limited by the size and shape of a product element. For example, tensile testing of the shaft of a motor is possible only if the shaft is long enough for such testing. However, there should be no problem measuring the hardness of the shaft.

It may be unrealistic for an organization to invest the time and the resources necessary to evaluate the reliability of the various critical product elements. Instead, design teams can get information about general reliability characteristics from suppliers and published technical references once the materials to be used have been identified.

Consider the skillet shown in Figure 8.3. An analysis of competitors' products should include an evaluation of the materials that constitute the handle, the pan base material and coating, and the strength of the handle and the base material of the pan. An evaluation of the pan's heat distribution performance might involve analysis of the temperature at different points within the pan. For reliability, the wear characteristics of the coating can be evaluated by moving a rough material,

FIGURE 8.3

Skillet.

like sandpaper, back and forth over the coating and measuring the number of back-and-forth cycles until the coating wears away.

8.2.3 Evaluate Customers' Interest in the Product Concept in the MRD

Evaluating customer interest in a product concept is primarily a marketing activity and will not be covered here.

8.2.4 Identify Risks and Risk-Mitigation Strategies

After completing the previous steps, a design team must reassess the risks associated with the product concept in the MRD. From the materials engineering perspective, the risks are associated with the uncertainty about the suitability and the cost of the materials and manufacturing processes, the available manufacturing capabilities, and the cost for materials selection and process development. Perhaps the analysis of competitors' products revealed unfamiliar materials or manufacturing processes that might have to be used to realize the desired performance and reliability in the new product. This will require some design team research into the materials and processes to better understand their complexity and cost.

Maybe customers expect greater performance and reliability than the design team expected, which may require the use of more expensive materials and manufacturing processes that will add cost to the product. The information obtained from assessing competitors' products will help designers better understand whether this is true. If there are no comparable products to assess, then determining whether more expensive materials and processes must be used may require continued development until there is enough design detail to identify materials and process options. This strategy may offer an unacceptable level of risk.

8.2.5 **Determine Whether There Are Unacceptable Risks**

The entire product team (i.e., marketing, sales, and engineering) must decide whether any of the risks identified in the previous step are unacceptable. If any are, then the team can either cancel the project or revise the marketing requirements document.

8.2.6 **Revise the MRD**

A design team may decide to revise the MRD based on the information obtained from the detailed market analysis. Perhaps there was not enough customer interest in the product, as defined in the marketing requirements document, or that based on the analysis of competitors' products greater functionality is required. If the MRD requires revision, then it will be necessary to repeat the product planning considerations discussed in Chapter 7.

The design team must identify the potential materials and manufacturing processes that can be used and the risks associated with them. If changes to the MRD would result in a product that does not satisfy the customer's wants and needs, then the project should be canceled.

Write Detailed Product Specification Goals

General Goals
- Develop a set of design requirements that is thorough and accurate.

Materials Engineering Goal
- Develop a set of design requirements that is thorough and accurate.

8.3 WRITE DETAILED PRODUCT SPECIFICATION

Once the design team has a list of the engineering requirements, it can start writing a detailed product specification. The information that goes into this document was discussed in Chapter 7.

The general goal and the materials engineering goal for this step of the process is to develop a set of requirements that is as thorough and accurate as possible. From a risk assessment perspective, a comprehensive product specification improves the probability of developing a product that meets the customer's wants and needs. From the materials engineering perspective, a comprehensive product specification increases the likelihood of selecting optimum materials in a timely manner.

> **Product Concept Goals**
>
> **General Goals**
>   - Generate product concepts that satisfy the product specification.
>   - Select product concepts that have a high probability of success.
>
> **Materials Engineering Goals**
>   - Identify potential materials and manufacturing processes that can be used for the product concepts.
>   - Identify the risks associated with using the potential materials and manufacturing processes.

8.4 PRODUCT CONCEPT GENERATION, EVALUATION, AND SELECTION

During concept generation, a design team creates product concepts based on the requirements in the product specification. These concepts are then evaluated, and the most promising are selected for further development.

Concept generation, evaluation, and selection occur at Type I and Type II companies. For a product designed by a Type II company, the design team generates concepts for a subassembly or component based on design requirements provided by a Type I company, which is the customer. Broader discussions of product concept generation, evaluation, and selection processes can be found in Pahl and Beitz (1996) and Ulrich and Eppinger (2004).

Concept generation, evaluation, and selection are part of an iterative process that consists of the following steps:

1. Generate concepts.
2. Evaluate concepts.
3. Determine whether any of the concepts are acceptable.
4. Select concept for further development.

The flowchart for this process is shown in Figure 8.1. Each of these steps is discussed from the materials engineering perspective later in this section.

8.4.1 Goals

There are two general goals of product concept generation, evaluation, and selection. The first is to generate design concepts that satisfy the product's design requirements. Whether or not a design team has identified all of the design requirements is another matter. The second is to select a product concept that has a high probability for success. From the design perspective, this means selecting a concept that can be developed into a product that satisfies all of the product specifications, meets the design schedule, and can be launched with few manufacturing problems. Thus, the selection of a concept design must include an evaluation of the technical, cost, schedule, and manufacturing risks for the design being considered.

A design team may decide to continue development of multiple concept designs that have different possibilities for success. The purpose is to determine whether it is possible to reduce the risk of designs that are perceived to offer a greater competitive advantage compared to other concepts with risks that are lower. Then, at some later point during the development process, the design team can reevaluate the risks of the different designs and select a single design for further development.

From the materials engineering perspective, there are two goals. The first is to identify potential materials and manufacturing processes that can be used for the product concepts. The second is to determine the risks associated with the options of materials and manufacturing processes that can be used for the different concepts. This assessment helps the design team to eliminate excessively risky designs before investing too much effort in them. Also, the information helps the team to select the designs to develop further.

8.4.2 Generate Product Concepts

Considering the materials engineering perspective when generating product concepts helps design teams to more quickly refine concepts and identify alternatives. Chapter 1 pointed out that including multiple perspectives is valuable for quickly narrowing the set of design options to only those that have a good chance of being successful. In addition, including alternative perspectives reveals options that are obvious to one perspective but not to the others.

During product concept generation, the engineering viewpoints that are typically represented are the mechanical or electrical engineering perspectives. Including the materials engineering perspective helps a design team to develop more feasible concepts. Even at this point in the product development process, it is possible to start identifying options for materials and manufacturing processes for the different concepts. With the materials engineering perspective represented the design team will have to consider engineering trade-offs at the beginning of concept development, rather than waiting until later in the design process when it is more difficult to make changes if suitable materials or manufacturing processes are not available. Also, including the materials engineering perspective improves the likelihood that concepts for which it will be difficult to find viable material and manufacturing options will be eliminated more quickly.

It may be possible to identify more than one material that can be used for a particular product element within a concept, with the materials offering a range of risk and capabilities. Ideally, at least one of the materials has little risk (or no risk). Higher risk materials might be included because they offer superior properties or cost less to use than the low-risk material.

In addition to helping to refine ideas for concepts, a materials engineer can put forward concepts based on using specific materials and processes that are unknown to the other design team members. The benefit of this is an increased number of possible design options.

8.4.3 Evaluate Product Concepts

Design teams evaluate the feasibility of product concepts based on customer acceptance and perceived technical and manufacturing risks. The risks are evaluated based on their estimated impact on the product development schedule, costs to produce the product, and ability to satisfy the design requirements. The concept evaluation process is different from organization to organization, and the approaches range from instinctive analysis to quantitative assessments and everything in between (Ulrich and Eppinger, 2004).

From the materials engineering perspective, the risks for any particular concept are associated with the likelihood of finding materials and manufacturing processes that enable the product to satisfy its design requirements. Low-risk concepts have options of materials that have a high probability of enabling the product to meet its design requirements and will not be difficult to form into the intended product element. Some evaluations may be required to verify that the materials are acceptable. In contrast, high-risk product concepts have options of materials for which less information is available, leading to more uncertainty about the suitability of the material and the ease of forming the material into the intended product element. Consequently, more evaluations will be required to verify that a higher-risk material is suitable.

Mitigating any material's related risks requires evaluating its properties and reliability. The cost, staffing, and time required for these evaluations depend on the amount of information available about the materials and the use conditions to which they will be exposed. As the number of higher-risk materials considered increases, the costs, effort, and time to verify the suitability of the materials also increase. Additionally, the cost and availability of the materials options should be investigated. Mitigating any risks related to manufacturing processes requires obtaining information about the ease of manufacturing a particular product element out of a particular material.

As a design team evaluates the product concepts, it must determine whether the concepts are feasible from the standpoint of customer acceptance and risk. If no concepts are acceptable, then the design team must decide from one of the following choices:

- Modify the concepts to reduce the risks.
- Generate new concepts.
- Revise the product requirements by (a) reducing the number of functions, (b) lowering the performance or reliability requirements, or (c) increasing the budget to make the product.
- Stop the project.

Revising the product requirements is acceptable as long as doing so does not result in a concept that fails to meet the customer's wants and needs. This is possible when a design team initially sets out to design a product with functions, performance, reliability, or a price that exceeds customer expectations. After revising the product specifications, the design team can either generate new design concepts or work with the ones that were previously generated.

One option for addressing the lack of acceptable concepts has not been discussed, and that is to continue forward with the design with the understanding that there is increased risk that the product will not satisfy all of the design requirements. It would seem that this option should never be a consideration. However, design teams implicitly select this option when they avoid making the decision altogether. That is, the design team knows that it is considering materials and manufacturing processes for which there are several significant unknowns and chooses to ignore consideration of the entire decision. In this case, the design team implicitly "chooses" to hope that the design will succeed.

8.4.4 Select Product Concepts

Once acceptable concepts have been identified, the design team selects the one that is believed to offer the greatest chance of success. Sometimes more than one is selected for further development so that more information can be obtained for the different concepts before selecting the one to complete.

REFERENCES

Pahl, G., and W. Beitz, *Engineering Design: A Systematic Approach*, Springer, 1996.

Ulrich, K. T., and S. D. Eppinger, *Product Design and Development*, 3rd edition, McGraw-Hill, 2004.

Materials Engineering Considerations for System-Level Design

9

9.1 INTRODUCTION

System-level design involves defining and developing design concepts for the subassemblies and product elements that make up the product. There is a general and a materials engineering goal for this design. The general goal is to develop a product design that has a high probability of satisfying the product specification and marketing requirements document (MRD). The materials engineering goal is to identify potential materials and manufacturing processes that will enable the design to satisfy the specification. The following are the system-level design steps:

1. Develop a plan for product options, product family, and set target sales price point(s).
2. Design the product architecture.
3. Design subassemblies and product elements.
4. Develop the sourcing strategy for materials, components, and subassemblies.
5. Select suppliers of subassemblies and components to be designed or manufactured by Type II and Type III suppliers.
6. Identify risks and strategies to mitigate risks.
7. Determine whether all of the risks are acceptable.

Figure 9.1 shows a flowchart for system-level design.

System-Level Design Goals

General Goal
- Develop a product that has a high probability of satisfying the product specification and marketing requirements document.

Materials Engineering Goal
- Identify materials and manufacturing process options that will enable the product to satisfy its specifications and marketing requirements document.

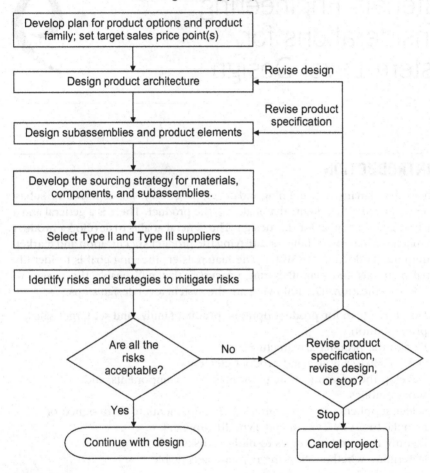

System-level design business and engineering perspective

Develop plan for product options and product family; set target sales price point(s)

↓

Design product architecture ← Revise design

↓

Design subassemblies and product elements ← Revise product specification

↓

Develop the sourcing strategy for materials, components, and subassemblies.

↓

Select Type II and Type III suppliers

↓

Identify risks and strategies to mitigate risks

↓

Are all the risks acceptable? → No → Revise product specification, revise design, or stop?

Yes ↓   Stop ↓

Continue with design   Cancel project

FIGURE 9.1

Flowchart for system-level design.

As the flowchart in Figure 9.1 shows, system-level design is an iterative process that involves design revisions or changes to the product specification. Design revisions require trade-offs that involve changes to the designs of various subassemblies and product elements to improve the likelihood that they will perform as required. One possible trade-off is modifying the physical construction of a product element to accommodate the use of a different material. Such changes are made if a design team realizes that it will be unable to complete development of a design concept that satisfies the product specification or that the schedule requirements are too aggressive.

Aspects of the system-level design process are similar to the concept generation and selection process. There will be concept designs for the subassemblies and product elements, which may evolve as more knowledge and information is obtained about the various designs and as unacceptable risks or opportunities for improvement are identified.

The decisions associated with the system-level design are applicable to Type I and Type II companies. Type II companies rely on a set of design requirements that the customer provides. After receiving these design requirements, a Type II company addresses the other decisions that follow in the flowchart shown in Figure 8.1. The system-level design phase is not applicable to Type III companies. As discussed in Chapter 1, Type III companies manufacture their products based on designs provided by their customers, which can be either Type I or Type II companies.

The remainder of this chapter discusses the materials engineering perspective for system-level design. Items 1 and 2 from the list presented earlier are not discussed because they do not require this perspective.

9.2 DESIGN SUBASSEMBLIES AND PRODUCT ELEMENTS

After a product concept has been selected, a design team begins the design of the major subassemblies that make up the product. Then, the subassemblies are divided into smaller subassemblies. Finally, the lowest-level subassemblies are divided into product elements. As this process progresses, the team defines the design requirements for the subassemblies and finally the product elements. Once these requirements have been defined, it is possible to identify the materials selection criteria for each product element.

The speed with which the materials selection criteria are compiled for each product element and the thoroughness and accuracy of the information depend on the team's level of knowledge about the requirements for the product and subassemblies. Incomplete information about these items impedes identification of the materials selection criteria because the design team will have to add the missing information while working on other aspects of product development.

An even worse consequence of missing or inaccurate selection criteria is identifying and selecting materials that will not enable a product's design requirements to be satisfied. In this situation, the product may not have the required performance, may not pass product verification testing, or may fail prematurely during customer use. If the product does not meet the performance requirements, the design team must decide whether or not to redesign it. If the product fails verification testing, then it must be redesigned and retested, which adds costs and delays the product launch.

Alternatively, the materials used may be more expensive than necessary because an engineer identified and selected materials with properties that exceed those required to achieve the desired product performance and reliability. As a result, either profit margins are reduced or the company must raise the product's sales price.

The process for developing concepts for subassemblies and their constituent product elements consists of the following steps:

1. Develop subassembly concepts.
2. Decide whether to purchase off-the-shelf subassemblies or design custom subassemblies.
3. Decide whether to design custom subassemblies internally or externally.
4. Develop product element concepts.
5. Decide whether to make or buy custom subassemblies and components.

The flowchart for this process is shown in Figure 9.2. Each step is described in more detail in the following sections.

9.2.1 Develop Subassembly Concepts

As discussed in Chapter 2, the highest-level subassemblies are designed directly from the product's design requirements, with the specific product functionality of each subassembly determined by the design team. The designs of lower-level subassemblies and product elements are based on the design requirements for the subassembly in which they are used. The functionality and the other design requirements influence the physical construction of each subassembly, which includes its shape, dimensions, components, lower-level subassemblies, joints, and in-process structures.

Lower-level subassemblies within the highest-level subassemblies can be identified after concepts for the highest-level subassemblies have been selected. This involves dividing the highest-level subassemblies into portions that can be most easily designed and manufactured. Some products may have only one level of subassemblies. Other products will have many levels of subassemblies.

As a design team develops the concepts for the subassemblies, it must also define their design requirements. Also, the design team should start to record the information in a specification. All of the specification information discussed in Chapter 7 must be included. At this point in the process, a great deal of information about the design requirements may be missing. This information will be added as the design is refined during the detail design phase.

Designing subassemblies and defining their design requirements is an iterative process that requires trade-offs and optimization to arrive at subassembly designs that satisfy the product's design requirements. In fact, changes to the designs can continue into the detail design phase.

9.2.2 Decide Whether to Purchase Off-the-Shelf Subassemblies or Design Custom Subassemblies

Off-the-shelf subassemblies are purchased from Type I companies. The process for selecting off-the-shelf subassemblies is discussed in the next chapter.

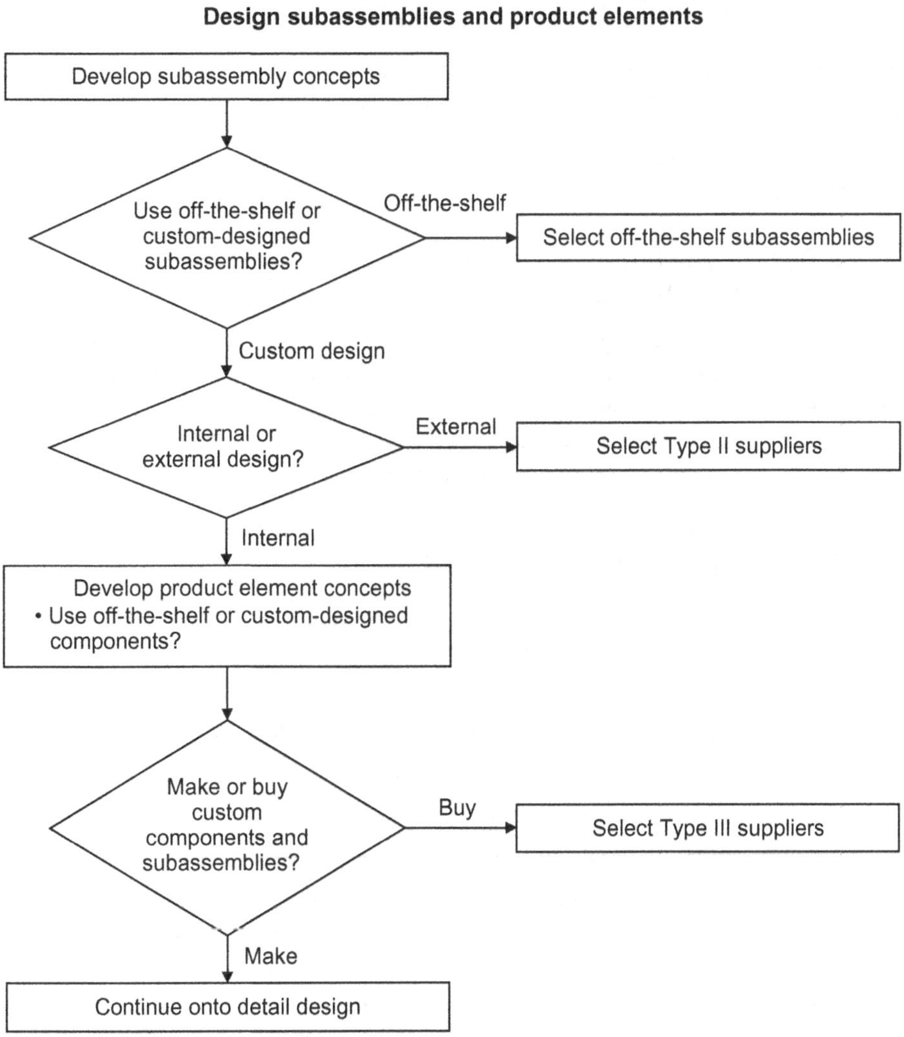

Design subassemblies and product elements

```
┌─────────────────────────────────┐
│  Develop subassembly concepts   │
└─────────────────────────────────┘
                 │
                 ▼
          ◇ Use off-the-shelf or        Off-the-shelf    ┌──────────────────────────────────┐
            custom-designed        ──────────────────▶   │ Select off-the-shelf subassemblies │
            subassemblies? ◇                             └──────────────────────────────────┘
                 │
           Custom design
                 │
                 ▼
          ◇ Internal or              External         ┌──────────────────────────┐
            external design? ◇   ──────────────────▶  │  Select Type II suppliers │
                 │                                     └──────────────────────────┘
              Internal
                 │
                 ▼
┌─────────────────────────────────────┐
│  Develop product element concepts   │
│ • Use off-the-shelf or custom-designed │
│   components?                        │
└─────────────────────────────────────┘
                 │
                 ▼
          ◇ Make or buy
            custom                   Buy           ┌───────────────────────────┐
            components and     ──────────────────▶ │  Select Type III suppliers │
            subassemblies? ◇                        └───────────────────────────┘
                 │
              Make
                 │
                 ▼
┌─────────────────────────────────┐
│   Continue onto detail design   │
└─────────────────────────────────┘
```

FIGURE 9.2

Flowchart for designing subassemblies and product elements.

9.2.3 **Decide Whether to Design Custom Subassemblies Internally or Externally**

Custom subassemblies can be designed by the company making the product or by another company (i.e., a Type II company). Typically, when a Type II company is used to design a subassembly, the same company also manufactures the subassembly. The process for selecting a Type II supplier is discussed later in this chapter.

9.2.4 Develop Product Element Concepts

The next section discusses the process for developing product element concepts. For components, the design team decides whether to purchase off-the-shelf components or design custom components. Off-the-shelf components are purchased from Type I companies. The process for selecting off-the-shelf components is discussed in the next chapter.

9.2.5 Decide Whether to Make or Buy Custom Subassemblies and Components

Custom subassemblies and components designed by the company making the product are either manufactured by the company or manufactured by and purchased from a Type III company. The focus here is on custom subassemblies and components manufactured by and purchased from a Type III company. The process for selecting a Type III supplier is discussed later in this chapter.

9.3 DESIGN PRODUCT ELEMENTS

After concepts for subassemblies have been selected, it is possible to design the product elements using the following steps:

1. Define subassembly design requirements.
2. Develop physical construction design concepts for the product elements.
3. Define product element design requirements.
4. Define materials selection criteria.
5. Identify potential materials and manufacturing processes.
6. Assess the risks of the design and develop strategies to mitigate the risks.
7. Decide whether the risks are acceptable. If not, then (1) continue with the design, (2) revise product element design concepts, subassembly design requirements, subassembly design concept, product design requirements, or product concept, (3) invent a new material or manufacturing process, or (4) cancel the project

Figure 9.3 shows a flowchart for this process. Steps 2 through 7 are discussed next.

9.3.1 Develop Physical Construction Design Concepts for Product Elements

Product elements can be defined using two approaches. The first is to develop design concepts for their physical construction, define their design requirements, and then identify materials that can be considered for use. This approach considers the mechanical and electrical design perspectives as being the principal

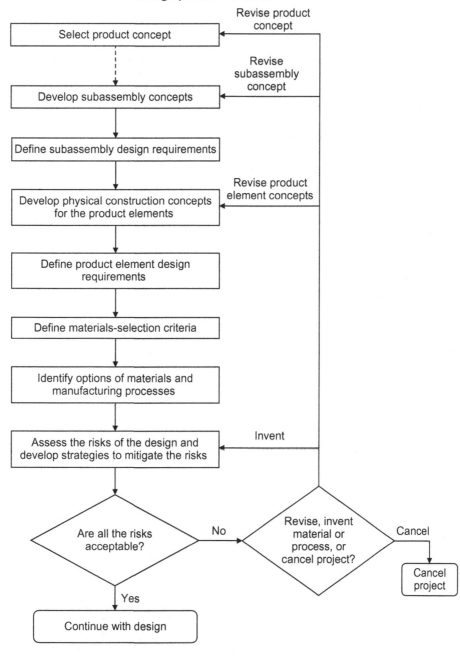

FIGURE 9.3

Flowchart for designing product elements.

concerns, which reduces the likelihood of developing optimized designs because the options for materials are constrained by the mechanical and electrical requirements.

A major drawback to this approach is finding out later, during detail design, that no materials exist that meet all of the design requirements for one or more product elements. Then the design team must decide whether to continue with suboptimum materials or redesign the product. Continuing with suboptimum materials runs the risk that the product will not meet all of its performance, reliability, or cost requirements. Redesigning the product may cause the schedule to slip and delay the product launch.

The other approach to defining product elements is to develop, evaluate, and modify their design concepts based on the options for materials that can be used, and select the design concepts that offer the lowest risk to finding suitable materials. This iterative approach helps prevent the problems described in the previous paragraph and increases the likelihood of developing a product that meets all of its design requirements.

9.3.2 Define Product Element Design Requirements and Materials Selection Criteria

Once concepts for the physical construction of the product elements have been developed, it is possible to define their design requirements. All of the requirements discussed in Chapter 2 must be defined for each product element before the potential materials are identified, because these design requirements are the materials selection criteria.

9.3.3 Identify Potential Materials and Manufacturing Processes

Once all of the materials selection criteria for a product element have been defined, the next step is to identify materials that can be considered for use. For a derivative product development effort, most or all of the materials will be the same as those used in previous versions of the product. However, it may be necessary to select some new materials if any of the design requirements are different compared to previous versions of the product. For a new product platform or fundamentally new product, the design team selects new materials for all of the elements in the product.

One approach to developing a list of potential materials involves the process of elimination, whereby information about materials is used to disqualify those that will not satisfy all of the selection criteria. For example, carbon steel would be rejected for use in a corrosive environment because it would quickly rust. This approach is used to eliminate classes of materials and then eliminate subclasses of materials and specific materials, leaving only the materials that appear feasible at that point in the process.

A class, subclass, or specific material is disqualified when it does not satisfy one or more selection criteria based on (1) data found in technical references, (2) experience, and (3) supplier technical data. The importance of conducting a thorough search for information cannot be overstated. Spending several hours reviewing published data and other information can have a significant impact if it disqualifies a material before too much effort is invested in further evaluations or, even worse, the material causes the product to fail product verification tests. Finally, another source of information about potential materials is the analysis of competitors' products.

To narrow down the number of potentially suitable classes of materials, the team determines the general classes that are known not to satisfy one or more of the selection criteria. Examples of classes of materials are shown in Table 9.1. A class of materials can be as broad as metals or polymers or one of the many subclasses within metals, polymers, ceramics, and composites. For example, within metals the subclasses include carbon steel, stainless steel, aluminum alloys, and brasses. For a particular product element, it may be possible to exclude all metals, polymers, ceramics, or composites from consideration or to exclude a subclass of materials within one of these groups of materials. For example, metals and ceramics might be excluded from consideration for a component that must have moderate strength and be lightweight and inexpensive.

The next step is to identify options of materials within the potentially acceptable classes and subclasses of materials. The same process of elimination as described earlier for disqualifying classes of materials can be used to disqualify subclasses of materials and specific materials within a subclass.

Determining the suitability of a particular class of materials requires access to various sources of information about materials and their features, properties, degradation characteristics, methods of manufacturing, ease of use in manufacturing, and cost. As discussed in Chapter 4, a great deal of published information is available in textbooks, handbooks, scientific and engineering journals, supplier data sheets, and industry standards. Finally, members of the design team may have experience with classes and subclasses of materials that were used for product elements in other products.

Table 9.1 Examples of Classes of Materials

Metals	Polymers	Ceramics	Composites
Carbon steel	Thermosets	Crystalline	Polymer matrix
High-strength, low-alloy steel	Thermoplastics	Oxide	Metal matrix
Stainless steels	Crystalline	Nonoxide	Ceramic matrix
Aluminum alloys	Amorphous	Glasses	Laminate
Copper alloys	Elastomers		
Brasses	Adhesives		
Bronzes	Epoxies		
	Silicones		
	Polyurethanes		

The technical data from one supplier of a specific material may indicate that the material is unsuitable for one reason or another, whereas the data for a similar material from a different supplier may indicate that the material has the potential to satisfy the selection criteria for the product element under consideration. This occurs when special formulations and additives are used in a material, and they influence the performance, reliability, and ease of use in the manufacturing process for forming the product element. Materials with special formulations and additives include plastic pellets for injection molding, adhesives, ceramic powders, solder paste, and braze compounds.

In *Materials Selection and Process in Mechanical Design* (Ashby, 2005), the process for identifying potentially suitable materials is discussed in more detail and many examples are presented. Materials selection charts are used to quantitatively illustrate the materials options available to satisfy two or more material properties requirements. An example is shown in Figure 9.4. These charts help design teams eliminate classes of materials that do not meet the design require-

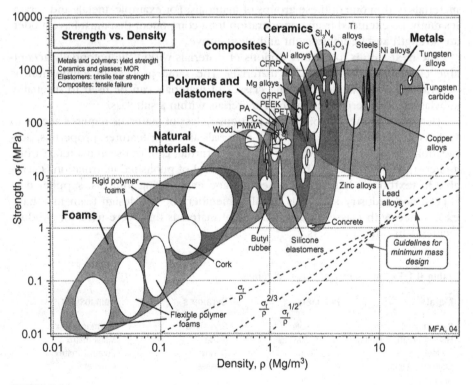

FIGURE 9.4

Materials selection chart for strength versus density. (*Source:* From Ashby, 2005; reprinted with the permission of Elsevier.)

ments and help to quickly determine whether a reasonable number of material options will be available for a particular product element.

A caveat to the process of elimination approach is that combinations of materials can be used that individually do not satisfy all the selection criteria but offer complimentary properties that do. For example, zinc-coated carbon steel can be used in a corrosive environment because the zinc provides galvanic corrosion protection to the steel. Using a combination of materials requires that each one satisfies specific selection criteria and that a manufacturing process is available for combining them.

As a design team considers materials, it will also have to consider manufacturing processes to use for forming each product element. Some materials must be used in conjunction with certain processes. Conversely, many processes require the use of specific materials or types of materials. For a combination of materials within a product element, the team must consider the compatibility of each material with the processes that follow, such as exposure to high temperatures or chemicals. Finally, the cost of prospective processes and the time and expertise required to develop the processes must be considered. The costs include those for process equipment, output evaluation equipment, and consumables. So, although a particular material may seem attractive, the processes required may not.

9.3.4 Assess Product Element Design Risks and Develop Strategies to Mitigate Them

From the materials engineering perspective, product element design risks arise from the following sources:

Insufficient materials information. There are few or no risks if the materials were used before in a similar application. It's another issue if the options of materials are new to the design team and there are a number of unknowns about them, even after researching published information. These risks are mitigated by evaluating the materials and performing materials reliability testing.

Inexperience with the manufacturing methods. It can be difficult to develop manufacturing processes using unfamiliar materials and manufacturing methods. These risks are mitigated through the process development, researching published technical resources, and discussions with suppliers of manufacturing equipment.

Availability and cost of materials. Using uncommon materials or specifying a narrow range of materials' features or properties will reduce the number of potential suppliers and might add to the costs. Mitigating these risks requires that the design team start researching the supply chain as soon as possible.

Availability and cost of off-the-shelf components and subassemblies. Restricting the use of certain substances for environmental reasons will reduce the number of potential suppliers and might add to the cost items. The same is true for requirements on the portions of a component or subassembly that will be secondary input materials. Mitigating these risks, again, requires researching the supply chain as soon as possible.

As information is obtained, a design team can reassess the suitability of a product element design and decide whether to change it. The discussion in the next section also applies.

9.3.5 Decide Whether the Risks Are Acceptable

After going through the process of identifying potentially suitable materials, it may be determined that no options exist or too few low-risk options exist for one or more product elements. The design team must then choose among the following options:

- Revise product element design concepts
- Revise the subassembly design requirements
- Revise the subassembly design concept
- Revise the product design requirements
- Revise the product concept
- Invent a new material and/or manufacturing process
- Cancel the project

To meet the product development schedule, it is desirable to face this decision as early as possible in the design cycle before spending too much time and money pursing a design that is not feasible.

Sourcing Strategy Development Goals

General Goal
- Categorize the items used within a product according to their relative importance and the risk of the available supply base for each item.

Materials Engineering Goals
- Identify the supplier materials and processing expertise required for each material, component, or subassembly to be used in the product.
- Identify materials that pose a supply risk and identify less risky materials.

9.4 DEVELOP SOURCING STRATEGY

Sourcing strategy development is one portion of the strategic sourcing process, and strategic sourcing is one element of supply chain management. Supply chain management is concerned with managing all the resources that go toward meeting the customer demand for a product once it is in production. The resources include the suppliers that provide goods and services used to create a product and the manufacturing facilities for making the product.

Supply chain management involves the following:

- Developing strategies for managing the resources
- Selecting suppliers
- Developing a set of pricing, delivery, and payment processes with suppliers
- Developing processes for managing the inventory of goods and services received from suppliers
- Scheduling the activities necessary for production, testing, packaging, and preparing the product for delivery

The goal of supply chain management is to optimize the use of the resources to maximize the probability of fulfilling customer demand for a product and minimize the costs to do so.

The purpose of strategic sourcing is to continuously improve and reevaluate a company's purchasing activities. The goals of strategic sourcing include reducing spending, managing supply risk, and complying with industry standards and government regulations. The steps in a strategic sourcing process are as follows (Toshihiro, 1994):

1. Assess a company's current spending to understand what is bought and from whom
2. Assess the supply market to understand who offers what products and services
3. Develop a sourcing strategy to determine from whom to purchase goods and services
4. Identify suitable suppliers who will minimize risks and costs
5. Negotiate with suppliers for products, services, and prices
6. Implement a new supply structure
7. Track results and restart the assessment for continuous improvement

The goal of sourcing strategy development is to select suppliers with services, materials, components, or subassemblies that help minimize the cost and the risks to make a successful product. A purchasing organization seeks to select suppliers based on the sourcing strategy and the strategic sourcing objectives. The suppliers can be other companies or an organization within the same company.

An organization's sourcing strategy has a large impact on the materials used within a product and on the control of the properties of those materials. A sourcing strategy influences the amount of engineering attention that a particular product element or subassembly receives during product development. It also influences the suppliers used to provide a service or make a material, component, or subassembly. More engineering resources are devoted to product elements and subassemblies that have a higher perceived value to and impact on the success of a product. This includes the attention paid to the selection of the materials that constitute the items.

9.4.1 Sourcing Strategy Considerations

Developing a sourcing strategy involves the following two considerations about the input materials, the components, and the subassemblies used within a product: (1) their relative importance within the product and (2) the risk of available supply for a particular item. The importance of an item is related to its impact on the performance and reliability of the product in which it is used. For example, the heater element in an industrial furnace has higher importance than the handle on the furnace door. Even though both are required for the furnace to meet the product requirements, the heater element enables performance and reliability capabilities that can give the furnace distinctive characteristics and give the company a competitive advantage over other furnace producers. A door handle will not enable performance attributes that give a competitive advantage.

The risk of available supply considers the number of suppliers that provide a service or sell an item, and it increases as the number of potential suppliers decreases. The risk of available supply depends on the following factors:

The technical complexity of the item. As the complexity of the item increases, the number of suppliers capable of producing the item decreases. This occurs because of the higher cost and technical competence required to make more complex items compared to less complex items. For example, the risk of available supply for an integrated circuit is much higher than for a fastener because there are many more fastener manufacturers than integrated circuit manufacturers. The cost and technical expertise is much greater to produce an integrated circuit than a fastener. Technical complexity is a relative term that depends on the product being considered. For example, within a particular product there may be different fasteners that have different complexities and supply risks depending on the function of each particular fastener.

The uniqueness of the item. As the uniqueness increases, the number of available suppliers decreases. The uniqueness is often a result of specifying a material, component, or subassembly that is produced by few suppliers.

The quantity needed. There can be two situations. For some items, as the desired quantity decreases, the number of potential suppliers may decrease because fewer suppliers are willing to provide the item in smaller quantities at a price that the purchasing company is willing to pay. Conversely, for other items, as the desired quantity increases, the number of potential suppliers may decrease because fewer of them have the capability to manufacture large quantities of an item.

9.4.2 Sourcing Strategy Matrix

The relative importance of each of the items used in a product and the risk of the available supply for each item can be used to separate items into the following categories:

- Routine
- Leverageable
- Bottleneck
- Strategic

Figure 9.5 shows the matrix format for displaying the relationship among these four categories and the two sourcing strategy considerations.

The low-importance/low-risk section is reserved for *routine* items that are required in a product but do not differentiate a company's product from its competitors' products. Routine items are not complex and are relatively easy to produce. Therefore, many companies are capable of making these items. Using an automobile as an example product, the interior panel knobs, outside door handle, seat belt, and seat belt buckle are routine items.

The high-importance/low-risk section is reserved for *leverageable* items that are of high importance to the functionality of the product and for which many potential suppliers exist. Leverageable items can be relatively complex and require significant expertise to produce. These items represent a considerable amount of money to the company and to potential suppliers. Thus, the company is able to exert some buying power when looking for suppliers. Examples of leverageable items in an automobile are the door panels and body paint.

Low-importance/high-risk items are referred to as *bottlenecks*. These items do not add much value to a product and are difficult to obtain. Managing the supply

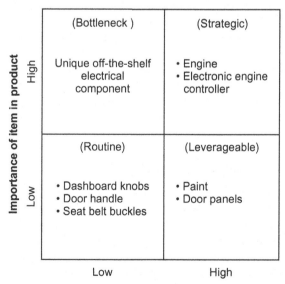

FIGURE 9.5

Sourcing strategy matrix.

of these items takes more effort than is warranted based on the item's value to the product. Items that fall into this category may have unique specifications or are in low demand. Low demand results in few suppliers. Sometimes companies select a material or off-the-shelf component that has unique characteristics that only one supplier can meet. This is a risky situation. One sourcing goal is to move items out of the bottleneck region. This can be done by substituting a different material, component, or subassembly that is more readily available than the current bottleneck item.

The final region of the matrix contains *strategic* items that have high importance and high risk. The high risk of available supply is due to the complexity or uniqueness of the item, which requires a high level of technical capability to produce. The difference between strategic and bottleneck items is that strategic items significantly differentiate a company's product from that of a competitor, add tremendous value to their products, and provide a competitive advantage. Consequently, it is worthwhile to invest time and money into the development of strategic items, and development of suppliers who have the technical expertise and manufacturing capabilities to produce the items. Examples of strategic items in an automobile are the engine and the electronic engine controller, which monitors and directs the engine functions to optimize engine performance.

As with most systems of categorization, the world is not a perfect 2×2 matrix with every item that a company purchases fitting neatly into each distinct category. Some items may be a combination of categories. Thus, the distinct categories represent extremes. The priority and attention any particular item receives depends on where it falls in the matrix relative to the other items. For example, an item may rank higher or lower in strategic significance compared to the other strategic items used in a product.

Developing a sourcing matrix requires input from engineers on the design team and people from the sourcing organization. The engineers understand the technical importance of each item within the product being designed. The people from the sourcing organization understand the supply base available for the different items and the costs to purchase them.

9.4.3 Benefits of Developing a Sourcing Strategy

The value of developing a sourcing strategy is twofold. First, it helps design teams prioritize how they should devote their resources, focusing their engineering efforts on the items that will give their product a competitive advantage and that may be the most challenging to develop and source. Also, there will be subassemblies, components, or materials for which it is not worthwhile to spend too much money or effort during product development. If it does not develop a sourcing strategy, the team runs the risk of devoting too much or too little attention and resources to the development of an item. For example, a design team might "intuitively" view an item as being routine and therefore treat it as such during development, even though an analysis would show that it is leverageable or even perhaps strategic.

The problem is that not enough attention will be given to the technical aspects of the item or to the capabilities of the supplier selected to provide it. A possible consequence is that the item does not perform as required or the supplier cannot meet quality requirements or delivery schedules.

The second benefit of developing a sourcing strategy is that it helps determine the expertise and capability required of suppliers. This influences the criteria used for supplier selection, perhaps eliminating options to use lower-cost suppliers who require significant oversight during product development and into production ramp-up. Alternatively, the information will indicate items for which a high level of technical expertise is not required, opening the potential to use lower-cost suppliers. However, it is still important to verify that lower-cost suppliers have the expertise to produce the item as required.

9.4.4 Materials Engineering Perspective of Sourcing Strategy

The sourcing strategy category assigned to an item influences the materials used in the item as a consequence of the resources allocated to develop it. The reality is that engineering organizations, which have finite resources, will not be able to put the same amount of effort and money into the engineering of every item. More resources will be allocated to engineer a strategic item than to engineer a routine item. The availability of resources affects the attention paid to the selection of the materials within an item and selection of the supplier. Therefore, the materials selected for a routine item stand less of a chance of being optimized than do the materials selected for a strategic item. Thus, it is important that each item is properly categorized so that the engineering effort, including materials selection, is proportional to the item's actual importance.

The cost of materials used in a product is also influenced by the sourcing strategy categorization. If a component or subassembly is identified as routine, then common, low-cost materials should be used wherever possible. For strategic items, it may be acceptable to invest in high-end or novel materials.

Proper identification of the sourcing category for each item also influences the control of the properties of the item's materials. There will be a different emphasis on finding a supplier capable of producing an item that consistently meets its performance and reliability requirements based on whether the item is strategic, leverageable, or routine. The reality is that more resources are put into finding and working with suppliers of strategic and leverageable items than with suppliers of routine items.

When strategic and leverageable items are incorrectly categorized as routine, the items do not receive the attention they require during product development, and the emphasis on selecting a capable supplier is lacking. This occurred with a product that required two plastic components to be ultrasonically welded together. The weld joint was viewed as being routine even though the product was new and the weld joint was a major part of the product's mechanical structure. The design team was inexperienced with ultrasonic weld joints, and little effort was put into understanding the weld requirements and the welding process. Also, little effort was put into evaluating the capability of the supplier selected to make the welded joints. As a result, the weld joints failed after the product was launched, which resulted in customer dissatisfaction. Also, a great deal of time was spent addressing the problem. This could have been avoided if the joint had been identified as being strategic, prompting the team to devote the proper resources to completely understand the joint's design requirements and to find the most appropriate supplier.

Finally, whenever an item is categorized as a bottleneck because of the materials used, the design team should be motivated to look for alternative materials that change it from a bottleneck item to a routine item.

9.4.5 Develop a Sourcing Strategy for Primary Input Materials, Components, and Subassemblies

The sourcing strategy categorization of a material, component, and subassembly used within a product depends on the following criteria:

1. Whether the item enables a critical performance feature of the product
2. Whether the item helps differentiate the product from other products
3. The design and manufacturing teams' experience with an item and the materials and processes for producing the item
4. The industry experience with the item and the materials and processes for producing the item
5. The number of suppliers that produce the item
6. For an off-the-shelf component or subassembly, the number of suppliers that produce the materials used in the item

The last item on the list requires that a design team have some understanding of the materials used in suppliers' products and the supply base for those materials. If a supplier of a component or subassembly uses bottleneck materials, then the item may become a problem for the design team if for some reason the materials are no longer available.

The sourcing strategy considerations for a product already in production or a derivative product in development differ from those for a fundamentally new product. For a product in production or a derivative product in development, the experience of the design and manufacturing teams with the materials and processes may enable items to be moved from strategic to leverageable and leverageable to routine.

This recategorization enables the use of lower-cost suppliers of primary input materials, components, and subassemblies. However, during development of a new platform product or fundamentally new product, there may be a lack of experience with the materials and manufacturing processes. This, combined with the pressure to meet product launch deadlines, may require categorizing more items as leverageable or strategic.

For new platforms and fundamentally new products, it is best to limit suppliers of high-importance items to those that have strong technical and manufacturing capabilities and have experience with the materials and processes used to make the item. Even though the cost per unit may be not be as low as desired, the benefits of working with more capable suppliers overshadow any risks associated with less capable suppliers. These benefits include fewer development problems and meeting development deadlines. It is less risky to develop a set of lower-cost suppliers for high-importance items once a product is in full production and all the details associated with designing and manufacturing the high-importance items are understood. Furthermore, the risk of available supply for any particular item within a product may change throughout its life as industry knowledge about the item increases and as more suppliers gain the necessary expertise and capability to produce it.

The familiarity of design and manufacturing teams with an item changes as they gain experience with the item and the materials and manufacturing processes used to produce it. Thus, a component or subassembly that was initially viewed as strategic may eventually come to be viewed as leverageable and perhaps, with enough time and experience, routine.

9.5 SELECT TYPE II AND TYPE III SUPPLIERS

Many Type I and Type II companies use Type II and Type III companies to design and manufacture the subassemblies and components that are used in their products. The process to use for selecting Type II and Type III suppliers of custom-designed and manufactured subassemblies and components consists of the following steps:

1. Write a statement of work.
2. Give the statement of work to prospective suppliers.
3. Wait for suppliers to provide proposals.
4. Review proposals.
5. Evaluate suppliers.
6. Select suppliers.

A flowchart for this process is shown in Figure 9.6. The first step is discussed in this section. There is not much to say about the second step. The supplier proposal process is discussed in the next section. Steps 4 to 6 are discussed in the next chapter.

9.5.1 Write Statement of Work

A statement of work is a document that communicates the requirements for a project. It is written by Type I and Type II companies to communicate the requirements for a component or subassembly to be custom designed and manufactured

Select Type II and Type III suppliers

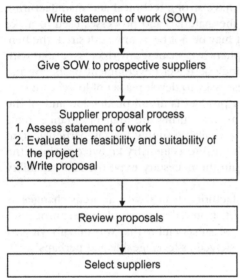

FIGURE 9.6

Flowchart for the supplier selection process.

by a Type II or manufactured by a Type III company. The contents of a statement of work are as follows.

Component or subassembly to be designed and manufactured by a Type II company

- Component or subassembly specification
- Packing and shipping requirements
- Development schedule requirements
- Production volumes and schedule requirements

Component to be manufactured by a Type III company

- Component specification
- Packing and shipping requirements
- Development schedule requirements
- Production volumes and schedule requirements

Subassembly to be manufactured by a Type III company

- Subassembly specification
- List of the components and primary input materials used to make the subassembly

- Specifications for the components and primary input materials used to make the subassembly
- Packing and shipping requirements
- Development schedule requirements
- Production volumes and schedule requirements

The development schedule requirements include the dates to receive prototype samples, perform product verification testing, develop manufacturing processes, and receive production-intent samples for testing. The product volumes and schedule requirements include information about the expected number of components or subassemblies that will be purchased at various points in the life of the design team's product.

Supplier Proposal Process Goals

General Goals
- Determine whether the project is a suitable fit.
- Estimate the resources, costs, and time required to design the item under consideration.
- Identify the risks associated with meeting all of the requirements in the statement of work.
- Write a successful proposal (if the project is a good fit and the risks are acceptable).

Materials Engineering Goals
- Identify options for materials and manufacturing processes that can be used.
- Determine the costs for materials selection and process development.
- Identify the risks associated with the options of materials and manufacturing processes.

9.6 SUPPLIER PROPOSAL PROCESS

A Type II company writes proposals to design or design and manufacture a component or subassembly for a Type I or Type II company based on the design requirements provided by the client company. A proposal is a document that describes the work to be performed by one company (contractor) for another company (client) and the fees that the client will pay for the work performed. A Type III company writes proposals to manufacture a component or subassembly for a Type I or Type II company based on a design that the client company provides.

There are four general goals for the proposal process. First, before writing a proposal, a company should determine whether the project is a good fit for its expertise and capabilities. If it is clearly not a good fit, then the project must be rejected. Second, a supplier must develop a reasonably close estimate of the resources, costs, and time required to design or manufacture the item under consideration. Excessive underestimation will lead to cost overruns and missed deadlines. Excessive overestimation may lead to proposed fees that are so high as to be noncompetitive.

The third goal is to identify the risks associated with being able to meet all of the requirements in the statement of work. The risks arise from two sources: (1) uncertainty about the information provided by the client and (2) uncertainty about the ability to assemble all of the required resources. The discussion of product planning in Section 7.2 applies to the considerations for the proposal process. Finally, the fourth goal is to write a successful proposal if it is determined that the project is a good fit and the risks are acceptable.

From the materials engineering perspective, three goals are associated with the proposal process. The first goal is to identify options of materials and manufacturing processes that can be used. The second goal is to estimate the required materials engineering effort and costs associated with materials selection and process development. The third goal is to identify the risks associated with the options of materials and manufacturing processes considered for use.

Assessing the project discussed in the client's statement of work is similar to the process applied during the planning phase for a Type I company. The steps for this review process are as follows:

1. Review the statement of work.
2. Validate the design requirements.
3. Determine whether the opportunity suits the company.
4. Determine whether the company has the resources to meet all performance, cost, and schedule requirements.
5. Determine whether all the risks associated with the project are acceptable.
6. Write and submit a proposal.

The review process steps are shown in Figure 9.7. Each of them is discussed in the next subsections.

9.6.1 Review Statement of Work

Type II and Type III companies must be able to identify all the resources required for a project and estimate with reasonable accuracy the time required to complete it. Additionally, the company must be able to assess whether or not it has the technical and manufacturing expertise and capabilities necessary for the project. From the materials engineering perspective, minimizing design and manufacturing risks requires the ability to identify potential low-risk materials and processes that can be used to produce the proposed component or subassembly.

All of this requires obtaining as much information as possible about the item's design requirements, which should be included in the statement of work. Just in

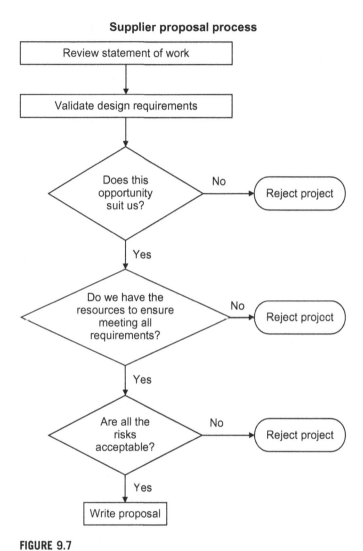

Supplier proposal process

FIGURE 9.7

Flowchart for supplier proposal process.

case some of the information is missing, or possibly incorrect, the supplier must validate the design requirements.

9.6.2 Validate the Design Requirements

It should not be assumed that the team from the client company has considered all of the constraints that affect the component or subassembly design requirements. Nor should Type II and III companies assume that all of the design decisions made by the potential customer are optimum or reasonable. In fact, it is not uncommon for component or subassembly requirements to be based on out-of-date information, misinformed decisions, and erroneous assumptions. When this occurs the contractor may then have to help solve integration problems resulting from faulty design requirements. Type II and Type III companies should make the effort to understand how their product will integrate into and be used in the client's product.

Engineering teams from Type II and Type III companies should validate the design requirements for the item under consideration; this involves the following:

- Confirming that the design requirements are complete and accurate and that none are missing, ambiguous, or incorrect
- Confirming that the product in which the item will be used will perform as required if the item satisfies its own design requirements

Design requirements validation is accomplished by reviewing each one with the client, discussing how the item will be used within the client's product, and identifying the conditions to which the item and product will be exposed. The following items should be validated:

Performance requirements. These include the functionality of the item and how a component or subassembly is joined to other parts of an assembly or subassembly.

Reliability requirements. This step involves ensuring that the customer has identified all of the relevant use conditions.

Verification tests. This step involves understanding if the verification tests are realistic or if any of the tests are difficult to pass and do not correlate well to the item's actual use conditions.

Production output evaluation requirements. These refer to any tests that evaluate the performance, materials properties, or materials features as the item is manufactured during production. It is important to understand these tests, the ease of performing them, and the equipment required.

A materials engineer should review the specification to determine the materials and processing options that exist to meet the item's design requirements. He or she can also help determine whether any of the design requirements in the customer specification are unclear, underspecified, or overspecified.

The validation process helps a supplier avoid spending time and money writing proposals for projects that would turn out to be problematic or unprofitable because they require more money, personnel, and time than the specification revealed. It also prevents the supplier from designing or making an item that meets the specification but does not perform as the client company intended (again because of unforeseen issues within a client's specification).

From the materials engineering perspective, the specification validation process improves the likelihood of getting all the facts required to make fully informed decisions about the materials and processes that can be used. The time spent in the validation process reduces the time needed to make decisions during development and improves the probability of a faster design process with fewer redesigns. If the design requirements for the item under consideration are not complete and accurate, then some material selection decisions will be misinformed or delayed. The first problem can lead to products that cost more than they should or an item that does not meet the expected performance and reliability requirements. Delayed material decisions lead to missed deadlines.

Sometimes it is not possible for companies to perform a thorough validation before writing a proposal. In that case, the company must decide whether it is willing to accept the risks associated with possibly not having all of the relevant information and whether it is willing to submit a proposal without all necessary information. The company can also include costs in the proposal that address the time and effort required to go through the validation process once the project begins, along with provisions that address the costs of the project in the event that the project is more complicated than the customer's original specification indicated.

On the other hand, if the engineers at the client company have the time to provide all of the necessary information after a proposal is accepted, then it seems that they should have the time to do this before accepting a proposal. In fact, it is risky to accept proposals based on engineering assumptions, especially if the item under consideration involves technologies, materials, and manufacturing processes that are new to both the client and proposing companies.

If the specification validation process is inadequate or skipped altogether, then difficult situations may arise for which the supplier is unprepared, leading to delays and higher costs. Ignorance is not bliss in this situation. It is not uncommon for suppliers to accept projects that initially involve unrealistic assumptions and predictions. Both the client and supplier hope that they will figure things out as they go along. The supplier encounters changes in product specifications as new information is obtained, and possibly finds out that certain expertise is required to evaluate design options and for manufacturing. When this happens, the supplier ends up spending a large amount of resources trying to make design and manufacturing decisions that are outside their technical expertise. Other problems include failed product verification tests because of the use of suboptimum materials and the item not performing as required in the customer's product. Fixing these problems is costly.

For example, a company won a contract to design and manufacture an electronics subassembly. During product development, it became clear that the client's design specifications for the subassembly's reliability requirements were incomplete and finding materials that satisfied the newly identified requirements would be difficult. Extra testing was required, and extra resources had to be devoted to solving the problem. In addition, special manufacturing processes were required to apply in-process materials. The subassembly manufacturer absorbed most of these additional costs.

9.6.3 Determine Whether the Opportunity Suits the Company

The considerations are similar to those discussed in Chapter 7 for a Type I company. As before, the company must decide whether or not the opportunity is a good fit for its engineering and manufacturing expertise and capabilities.

9.6.4 Determine Whether the Company Has Enough Resources to Meet All Performance, Cost, and Schedule Requirements

The contractor company must consider all of the risks associated with the project and decide whether to submit a proposal. Ultimately, the risks are related to the company's ability to produce the item, make a reasonable profit, and meet the design and production schedules. The perceived risks are affected by the thoroughness and accuracy of the design requirements, the gap between the technical expertise required and that available, and the gap between the engineering and manufacturing resources required and available. The risks are impacted by the materials engineering expertise and resources required for materials selection and process development. After assessing the project and its risks, a company must decide whether the business to be gained is worthwhile.

The design and manufacturing teams for the contractor must be able to honestly and accurately estimate the technical expertise, engineering and manufacturing capabilities, and personnel required for product and manufacturing process development. This is especially important when the design or manufacturing involves materials or technology with which a company has no experience. Many companies have taken on projects to design or manufacture a component or subassembly that ended up costing more money than it earned because the company underestimated the technical expertise, engineering resources, and manufacturing capabilities required for success.

9.6.5 Determine Whether All Risks Associated with Project Are Acceptable

Some of the major risks to Type II and Type III companies are associated with the following:

- Being able to determine whether all design and manufacturing requirements supplied by the client are thorough and accurate
- Being able to identify and select materials and manufacturing processes that enable the component or subassembly to meet all design requirements
- Being able to develop the item and capable manufacturing processes according to the design schedule

The first item is mitigated through careful validation of the statement of work. The second and third items are mitigated by the validation process combined with an honest assessment of the company's engineering expertise, manufacturing capabilities, manufacturing competence, and available resources. Obviously the risks are high when a company takes on a project that is far outside of its expertise.

9.6.6 Write and Submit a Proposal

A proposal from a Type II company will include the design fee, cost per unit of the item to be manufactured, project schedule, and some or all of the following information:

- The design concept
- Materials and manufacturing processes that will be used to make the product
- Costs for manufacturing process development
- Costs for manufacturing equipment and fixtures (also referred to as tooling)
- Risk assessment with respect to the development schedule and costs

A Type III company's proposal will include the cost per unit of the item to be manufactured, the project schedule, and some or all of the following information:

- Costs for process development
- Costs for manufacturing equipment and fixtures (also referred to as tooling)
- Risk assessment with respect to the development schedule and manufacturing costs

Additionally, a proposal can contain discussions of any of the risks and the strategies that will be used to mitigate them.

REFERENCES

Ashby, M. F., *Materials Selection in Mechanical Design*, 3rd edition, Butterworth-Heinemann, 2005.

Toshihiro, Nishiguchi, *Strategic Industrial Sourcing*, Oxford University, 1994.

Detail Design and Testing

10.1 INTRODUCTION

The detail design phase involves completing the product's design. The design team works toward completion of the specifications for the product and its subassemblies, product elements, and manufacturing processes. Like the other phases of product development, detail design is an iterative process. Design trade-offs are made as the design team learns more about the impact of design decisions on the performance, reliability, and cost of the product. Furthermore, the design team obtains more information about how to manufacture components, subassemblies, and the product, allowing it to modify and refine the designs and, if necessary, the product's design requirements.

The following elements of the detail design phase require the materials engineering perspective:

- Selecting suppliers of custom subassemblies and components
- Selecting off-the-shelf subassemblies and components
- Completing the design of custom subassemblies and components
- Developing manufacturing processes
- Completing subassembly, component, materials, and manufacturing process specifications
- Conducting product verification tests

The flowchart for this process is shown in Figure 10.1. The rest of this chapter discusses the material engineering considerations for each of these steps. Additionally, root cause analysis for problems that may arise is discussed at the end of this chapter.

Detail design process

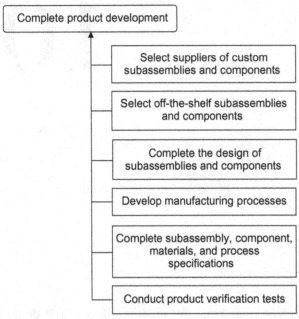

FIGURE 10.1

Flowchart for the detail design process.

> ### Selecting Custom Subassemblies and Components Suppliers Goals
>
> **General Goal**
> - Select suppliers that can design or manufacture custom subassemblies and components that minimize the total risk and cost associated with using the items in the design team's product.
>
> **Materials Engineering Goals**
> - Identify the risks associated with using the materials and processes recognized in the proposals.
> - Identify which suppliers have adequate experience with these materials and processes.

10.2 SELECT CUSTOM SUBASSEMBLIES AND COMPONENTS SUPPLIERS

Once the prospective suppliers have submitted their proposals, the design team must review them and assess the risks associated with each supplier's technical approach, engineering and manufacturing capabilities, and other criteria. From the materials engineering perspective, the risks are associated with the feasibility of the materials and manufacturing processes indicated in the proposals, the suppliers' experience and expertise with these materials and manufacturing processes, and the suppliers' manufacturing competence. The risks are assessed against each supplier's fee to design or manufacture the item. The client company selects the supplier that demonstrates a combination of low risk, a competitive fee, and a high probability of consistently meeting the design requirements.

From the materials engineering perspective there are two goals for supplier selection. The first is to determine the risks associated with using the materials and manufacturing processes identified in the proposals. The second is to determine whether a supplier has adequate experience with these materials and processes. This goal relates to controlling the variation of the properties of the materials to be used in the proposed item.

The supplier selection process consists of the following steps:

1. Review each proposal.
2. Assess the risks associated with each proposal.
3. (a) Assess each supplier's capabilities.
   (b) Evaluate samples that are similar to the component or subassembly being considered.
4. Determine whether the suppliers meet the minimum acceptance criteria.
5. Select a supplier.

A flowchart for the supplier selection process is shown in Figure 10.2. Each step is discussed next.

10.2.1 Review Each Proposal

As a design team reviews the proposals, it looks for certain information, including the following:

■ The supplier's approach to designing and making the component or subassembly
■ The costs associated with developing the item
■ The costs associated with process development
■ The item's cost per unit

Enough details should be available about the materials and manufacturing processes for the team to assess the feasibility and risks of a supplier's engineering approach.

10.2.2 Assess Risks of Proposal

A design team must determine the risks associated with the item meeting its design requirements. These risks are related to the proposed design and manufacturing strategy and are associated with, among other things, the materials and manufacturing processes indicated in a proposal.

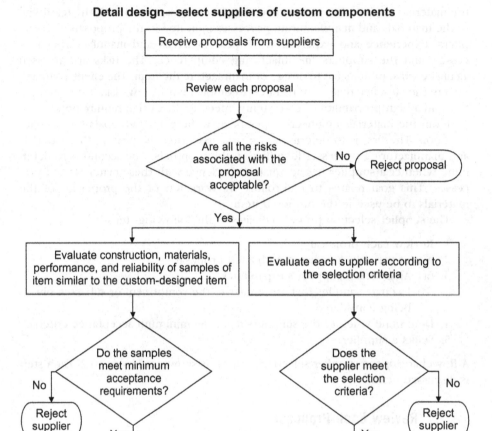

Detail design—select suppliers of custom components

FIGURE 10.2

Flowchart for selecting suppliers of custom components and subassemblies.

For an item to be custom designed and manufactured by a Type II supplier, the risk assessment includes considerations similar to those used to identify materials during concept development. In fact, the risks should be discussed in the proposal, along with a discussion of strategies to mitigate the risks and the supplier's experience with the proposed materials and manufacturing methods. The proposal should also explain the rationale for selecting the materials and processes. The discussion should include supporting materials performance and reliability data found in published literature.

Another consideration is a supplier's experience with the materials and manufacturing processes it has proposed. This applies to both Type II and Type III products. The proposal should include examples of products that the supplier previously built using the proposed materials and processes. It is important to validate that the performance and reliability requirements of the example products are similar to those of the item under consideration. If there are any significant differences, then the team must determine whether the requirements for the new design are more or less aggressive compared to the past designs. Finally, information about the performance, reliability, and manufacturing yields of the referenced products should be presented.

The proposal should also mention whether or not the materials comply with industry standards. If they do not, the rationale for using nonstandard materials should be explained.

Finally, the design team must assess whether the supplier is realistic about the costs to make the item. Will the supplier make a profit or ask for price increases?

10.2.3 Evaluate Supplier Background and Capabilities

A design team must decide whether or not it believes that a supplier can deliver the item as described and according to the product development and manufacturing schedule given in the proposal. The team also must assess the risks associated with using a particular supplier, as well as the supplier's fees.

The criteria for supplier evaluation and selection, in no particular order, are as follows:

- Technical expertise and capabilities
- Manufacturing capabilities
- Manufacturing competence
- Ability to provide engineering consulting
- Geographic location
- Company size and financial stability

Technical expertise and capabilities. Is the supplier knowledgeable about the materials to be used in the subassembly or component? Does the supplier have experience with the materials and processes indicated in the proposal? Does the supplier have the technical expertise to make good engineering decisions and resolve design and manufacturing problems?

Manufacturing capabilities. This relates to the supplier's manufacturing capacity, the different manufacturing processes available, and the capabilities of the manufacturing equipment. The ability to produce a high volume of the item, the capability of fabricating very large or very small components, or the ability to perform different primary and secondary processes are just a few of the many considerations when assessing manufacturing capabilities. These capabilities may extend to a supplier's manufacturing partners, who enable the supplier to extend its range of manufacturing processes and services.

Manufacturing competence. Is a supplier capable of controlling the manufacturing process inputs to make a product that consistently meets its design requirements? Is the supplier's manufacturing experience directly applicable to the item under consideration?

Ability to provide engineering consulting. A design team may determine that it does not have all of the expertise to make certain decisions about the design of its product and may want outside expertise. Many Type II and Type III companies have a great deal of technical expertise. In fact, an experienced supplier will have in-depth knowledge in its field of expertise combined with experience working on a wide range of products for different customers. Such suppliers have the knowledge and perspective that enables them to offer design suggestions that will have a positive impact on the materials selected and make it easier to form product elements that consistently meet their design requirements.

Geographic location. Is it important for the supplier to be located nearby? Sometimes proximity makes it easier for the team to communicate with the supplier or visit the supplier's location when necessary, and it contributes to the speed of delivery. Or is it acceptable for the supplier to be on the other side of the world? This decision affects the team's ability to assist with the supplier's product and process development and with the product quality problems that arise during production.

Company size and financial stability. All other things being equal, the size of a company and its financial stability indicate the resources that can be brought to bear to make decisions and solve problems. Also, for long-term projects, the financial stability indicates the likelihood that a company will be able to meet its obligations throughout the course of the project.

To evaluate manufacturing competence and capabilities, the team must visit the supplier to observe the manufacturing processes and to understand the technical capabilities for product development, controlling processes, and addressing problems. Although quality systems certifications such as ISO 9000 are important, they may not give a complete picture of a supplier's manufacturing and engineering capabilities.

The relative importance of each criterion depends on the sourcing strategy for the item under consideration. For an item that has been categorized as routine, cost may be more important than geographic location or ability to provide engineering consulting. For an item that has been categorized as strategic, a supplier's technical expertise, manufacturing competence, and ability to provide engineering consulting are of high importance.

Another factor to consider is a design team's experience with the item under consideration and with the materials and processes used to make it. If there is not much experience and the item is strategic or leverageable, then it is important to consider suppliers with strong technical expertise and a willingness to provide engineering consultation. However, if the design team does have experience with the item, then it may not need the engineering consultation. In fact, the design team can leverage its expertise to select a Type III supplier that is less expensive and less capable, but willing to accept guidance from the design team to help it develop more capable manufacturing (i.e., reduced output variation) processes or to increase its manufacturing capabilities.

To obtain materials, components, or subassemblies that consistently satisfy the performance and reliability requirements, it is better to have overly stringent selection criteria in regard to a supplier's technical capabilities and manufacturing or assembly competence. This helps prevent quality problems with the item being purchased. In contrast, supplier selection criteria that undervalue a company's technical expertise and manufacturing competence may lead to a lower per unit price for the item but may also lead to quality problems along with their associated costs.

10.2.4 **Evaluate Samples of a Comparable Item**

Design teams should verify that a prospective supplier is capable of manufacturing the item under consideration so that it meets the design requirements. It is possible to obtain this information by evaluating a similar item that the prospective supplier already produces. This verification process involves analyzing the critical product elements or materials within the comparison item to determine whether their material features and properties are acceptable and to ascertain the types and amounts of defects present. Any problems uncovered by the analysis will likely be present in the component or subassembly to be made for the design team.

Money, time, and effort must be spent to perform this analysis. However, it is a relatively inexpensive way to mitigate the risks of selecting a supplier that is not capable of properly designing and manufacturing the component or subassembly. Furthermore, the analysis is much less expensive than missed deadlines or low process yields related to poor supplier quality.

This analysis goes much deeper than evaluating the quality systems to verify that a supplier uses accepted methods to control its manufacturing processes. It also helps determine whether the supplier has the necessary technical expertise, including an understanding of the interactions between the process input and output variables. The team can learn all this early in the detail design phase, long before building any production tools.

The first step is to obtain samples of the supplier's comparison product. The product selected must be made using materials and manufacturing processes that are similar to those that will be used for the design team's component or subassembly. If a supplier is unable to provide samples, then the design team should question the supplier's capability of producing the item in question.

The specific analyses to perform depend on the sample and its intended use within the product. The following are descriptions of the general types of samples and the corresponding evaluations:

Component. Evaluate the relevant material properties, composition, microstructure, and defects. Also, evaluate the surface condition if the component is used in a joint or in-process structure. If the component is used in a joint, then assess the ability to form a good joint to the component. If the component is used in an in-process structure, then verify that the in-process material will adequately adhere to the component surface.

Subassembly. Evaluate the composition, microstructure, defects, and material properties of joints and in-process structures. If the subassembly will be joined to another subassembly, then the components (as secondary input materials) involved in the joints should be assessed for their composition, microstructure, and defects. Also, these components' surface condition and the ability to form good joints to them should be evaluated.

10.2.5 Select Supplier

A supplier should be selected based on its ability to minimize the total cost of its product. This goes beyond fees indicated in the supplier's proposal. The total cost also includes the costs of poor quality, which takes into account the costs that arise if the supplier's product does not satisfy its design requirements. The costs of poor supplier quality include the costs associated with failed product tests, low manufacturing yields, poor product reliability, and product recalls. These unpredicted costs can quickly eliminate any savings associated with using an incapable, low-cost supplier in place of a more expensive supplier with more technical expertise and manufacturing competence.

10.2.6 Supplier Development

In some situations, a design team will consider using a supplier whose engineering or manufacturing expertise and competence is not as strong as required. This occurs when a supplier offers a substantial technical or cost advantage compared to other suppliers. In this situation, many design teams recognize that they will have to invest resources to help the supplier improve. Whether it is worthwhile to invest the time, money, and effort is based on the total cost of the supplier's product or uniqueness of the supplier's product versus the probability of success and the size of the investment.

From the materials engineering perspective, problems with the supplier's product may originate with (1) using suboptimum materials and processes; (2) using unreliable manufacturing equipment, output measurement equipment, and/or output measurement processes; and (3) having inadequate control of the manufacturing process inputs.

A supplier may need help to improve the design of its product so that it performs better and is more reliable. Making improvements may mean using different materials or suppliers of materials. Also, it may be necessary to help a supplier improve its manufacturing processes in order to make a product that consistently meets the performance and reliability requirements.

For example, a supplier may have the ability to produce electrical resistors with a specific thermal coefficient of resistance that makes its product very attractive compared to resistors from other suppliers. However, the supplier may not have the ability to control the variation of the thermal coefficient of resistance. Consequently, many resistors do not meet the design requirements of the products in which they are used. The design team can help the supplier better understand how to control its manufacturing processes so that the variations are reduced to an acceptable level.

Selecting Off-the-Shelf Subassemblies and Components Goals

General Goal
- Select components and subassemblies that consistently meet the design requirements of the product being developed.

Materials Engineering Goals
- Select components and subassemblies made of materials that enable the product to meet its design requirements.
- Select components and subassemblies that can be easily joined to other components and subassemblies made of materials that enable the formation of joints that consistently meet performance and reliability requirements.
- Select components and subassemblies that are compatible with in-process materials.

10.3 SELECT OFF-THE-SHELF SUBASSEMBLIES AND COMPONENTS

Off-the-shelf subassemblies and components are made by Type I companies and used in the other companies' products. Examples of off-the-shelf subassemblies include motors, pumps, valves, electronic displays, and computer hard drives. Examples of off-the-shelf components include fasteners, gaskets, electrical resistors and capacitors, and metal, plastic, and ceramic tubing. When selecting off-the-shelf subassemblies and components, the general goal of a design team is to choose items that consistently meet the product's design requirements.

Several materials engineering goals are associated with selecting off-the-shelf items. The first goal is to select subassemblies and components made of materials that enable the product being developed to meet its design requirements. The design team must verify that any material degradation that occurs during use will not prevent the subassembly or component from meeting these requirements. In some cases, the manufacturer of a component or subassembly will have conducted reliability tests on its product and can provide test data that demonstrate the product's reliability. In other cases, it may be necessary to perform verification tests on a supplier's product. Finally, the design team should verify that the materials meet all industry standards and government regulations.

The second goal is to select subassemblies and components that can be easily and consistently joined to other components and subassemblies in the product being developed. A design team selects more than just a mechanical or electrical device when selecting a subassembly or component. It is also selecting secondary input materials to which a proper and reliable joint can be easily formed. These materials should be evaluated for their material features and whether or not a good solder, braze, weld, or adhesive joint can be made to the material. Also, they may have to be compatible with specific joint filler materials and joining processes already in use within a manufacturing production line.

The third goal is to select subassemblies and components made of materials that are compatible with in-process materials that will be deposited over the subassemblies and components. The surfaces that will be in contact with the in-process material should be evaluated for their material features and whether or not the in-process material adheres to the surfaces. Examples of surface materials features to evaluate are composition, cleanliness, and roughness.

The process for selecting off-the-shelf components and subassemblies is similar to that used for selecting suppliers of custom components and subassemblies and consists of the following steps:

1. Identify off-the-shelf items of interest.
2. (a) Obtain samples and evaluate their construction, materials, performance, reliability, and use in manufacturing processes.
   (b) Assess the supplier's capabilities and background.
3. (a) Decide whether the samples meet the minimum acceptance requirements.
   (b) Decide whether the supplier meets the selection criteria.
4. Approve or reject the item for use.

The flowchart for this process is shown in Figure 10.3. Notice that the two parts of steps 2 and 3 are conducted in parallel.

10.3.1 Identify Off-the-Shelf Items of Interest

Any off-the-shelf item being considered must satisfy all the design requirements, including those related to the materials that make up the item. Certain items are eliminated from consideration because they do not have the right performance, reliability, size, shape, or cost. Also, information about the materials can be used to eliminate certain items from consideration. Components and subassemblies that

Detail design—qualify off-the-shelf components

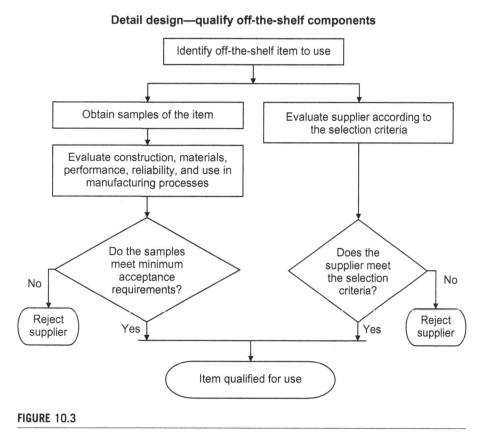

FIGURE 10.3

Flowchart for selecting an off-the-shelf component or subassembly.

do not meet required government regulations or industry standards for the materials must be rejected. Items that have secondary input materials that are incompatible with joint or in-process structure formation processes must be rejected.

Once subassemblies or components have been identified, the design team can move on to the next steps of the process. If no options exist for a particular component or subassembly, then the design team will have to decide whether to revise its design requirements or to redesign the product.

10.3.2 Evaluate Samples of the Item

The discussion in Section 10.4.2 for the evaluation of samples from a supplier of custom subassemblies or components applies here. There is one difference: Instead of using a comparable item, samples of the actual item being considered are evaluated.

10.3.3 **Evaluate Supplier Background and Capabilities**

The criteria for assessing the supplier of an off-the-shelf item are similar to those for selecting a supplier of a custom item. The sourcing strategy helps the team to determine the relative importance of the different criteria.

10.3.4 **Approve or Reject the Item for Use**

After the evaluations of the samples and the suppliers are complete, the design team decides whether to reject or approve a supplier's product for use. Rejection is most clear cut when the samples do not satisfy the design requirements for use in the design team's product. Rejection based on a supplier's background and capabilities is obviously more subjective. From the materials engineering perspective, a supplier's product should probably be rejected if the assessment of the product or the supplier's manufacturing competence indicates that the variations in the properties of the materials will be excessive.

Select Materials Goals

General Goals
- Select materials that will enable the product to meet its design requirements.
- Select materials that will enable product elements to be fabricated easily.

Materials Engineering Goals
- Select materials that will enable the product to meet its design requirements.
- Write specifications for the materials.

10.4 **SELECT MATERIALS**

Chapter 9 discussed the process for identifying options of materials. This section covers the evaluations required for selecting the materials to be used in a product element. The general and materials engineering goals of the materials selection process are the same.

The first goal is to select materials with properties that enable the product to satisfy its design requirements. The second goal is to select materials that enable product elements to be fabricated easily, resulting in a capable manufacturing process. This includes selecting off-the-shelf components and subassemblies made of materials that enable reliable joints and in-process structures to be easily formed.

The selection of the materials to use for a product element involves the following steps:

1. Determine the materials selection criteria based on the product element's design requirements
2. Identify materials that have the potential to satisfy the materials selection criteria
3. Evaluate the materials
4. Assess whether any of the materials satisfy all the materials selection criteria
5. Select the materials
6. Write materials specifications

A flowchart for this process is shown in Figure 10.4. The first two steps were discussed in the previous chapter. The remaining steps are discussed here.

After identifying options for the materials that can be considered, which was discussed in Chapter 9, the next step in the materials selection process is to evaluate the different materials. The evaluations should be performed (1) on the primary input materials that will be used to form components, fill joints, and form in-process structures and (2) on the secondary input materials of components and subassemblies that will be incorporated in a joint or in-process structure.

As discussed in Chapter 3, the candidate materials must be evaluated for the following:

- Material properties and their variations
- Material features and their variations
- Defects and their variations
- Ease of use in the manufacturing process
- Degradation and reliability characteristics
- Total cost to use

Each of these items was introduced in Chapter 3 and all except total cost will be discussed further.

Manufacturing process selection should also be considered during the materials selection process. Materials and product manufacturing processes are interdependent, and not all materials options are suitable for all manufacturing processes. For example, as discussed in Chapter 4 for a single aluminum alloy, the different casting processes resulted in different microstructures and properties. Also, requiring the use of a particular process may limit the types of materials that can be used, hence the manufacturing process design requirement discussed in Chapter 2.

The interdependence between the materials and the manufacturing processes requires that information from process development be considered when evaluating the feasibility of a material. This includes estimating how capable a process will be and the yields for a particular combination of materials and process.

Finally, materials with similar nominal compositions made by different suppliers should be considered as different materials. It should not be assumed that they

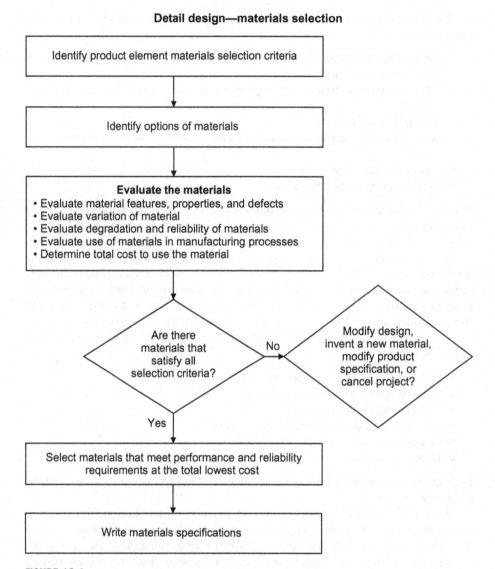

Detail design—materials selection

Identify product element materials selection criteria

Identify options of materials

Evaluate the materials
- Evaluate material features, properties, and defects
- Evaluate variation of material
- Evaluate degradation and reliability of materials
- Evaluate use of materials in manufacturing processes
- Determine total cost to use the material

Are there materials that satisfy all selection criteria?

No

Modify design, invent a new material, modify product specification, or cancel project?

Yes

Select materials that meet performance and reliability requirements at the total lowest cost

Write materials specifications

FIGURE 10.4

Flowchart for the materials selection process.

will have the same material features, properties, or defects. This applies to primary and secondary input materials. For example, for primary input materials, such as ceramic powders, adhesives, or metal bars, differences in the microscopic features or additives will give rise to differences in the ease of forming product elements that consistently meet their design requirements. Differences in the type, size, and number of defects present in the materials will result in differences in the reliability of the product element formed. Finally, there will probably be differences in the amount of variation in the material features and properties for materials produced by different suppliers.

10.4.1 Evaluate Material Features, Properties, and Defects

The specific features, properties, and defects to evaluate are dictated by the materials selection criteria for each particular product element. Because the information in published literature may be incomplete, the evaluations are used to provide the missing information.

10.4.2 Evaluate Variation in Material Features, Properties, and Defects

The material features, properties, and defects of primary and secondary input materials will vary from sample to sample and from batch to batch. A snapshot of the amount of variation of input materials can be obtained by evaluating the materials from different batches from a particular supplier.

As discussed in Chapter 5, the variation in the materials affects the capability of the manufacturing process being used to form a product element. The amount of variation for a particular material differs from supplier to supplier because they have different control over their manufacturing processes. In fact, the amount of variation for the assorted products of a particular supplier can also differ. Finally, the amount of variation depends on the type of material being considered. Some materials are inherently more difficult to produce than others, and there will be extra cost to exert tighter control over the manufacturing processes for the more difficult materials.

Variation of the input materials affects ease of manufacturing and whether a process is capable because the ability to manipulate the materials depends on their features, properties, and defects. For example, the molecular weight of a plastic affects its ability to flow into a mold during injection molding. The amount of lead in certain steel alloys affects their ease of machining. The viscosity of an adhesive affects the amount of adhesive applied by an automated dispenser. Understanding the variation of the input materials will help select materials that will enable capable manufacturing processes. The effects of input material variations on a product element and on the capability of a manufacturing process are evaluated during process development, which is discussed in Section 10.5.

10.4.3 Ease of Use in Manufacturing Processes

Ease of use refers to the ability to manipulate a material into the intended product element shape and dimensions and with the desired features, properties, and defects. The easier it is to manipulate a material, the faster a production line can manufacture its product. Some materials are easier to manipulate than others.

Also, the ease of manipulating similar materials from different suppliers is often very different.

The ease of using a particular material from a particular supplier is evaluated during process development. Prototype structures can be made that simulate the product element of interest. For example, prototype structures for solder joints or adhesive joints can be developed to evaluate the ease of depositing the desired amount of the material.

10.4.4 Evaluate Degradation and Reliability of Materials

The published information reviewed when identifying options may have indicated no known concerns regarding the reliability of a material. However, more information is usually required for the following reasons:

- The data available do not cover all of the conditions to which the materials will be exposed.
- The data available are for the general exposure conditions but do not cover the specific use conditions to which a product element will be exposed.
- The data available are for the general reliability of a material when exposed to the use conditions and do not cover specific applications where the geometry of product elements is a factor in their response to the use conditions.

Therefore, evaluation of the degradation behavior of the materials is often required to identify those that meet the reliability requirements. The considerations for the testing that is required are discussed in Chapter 6.

After testing is complete, the data can be plotted on a graph and analyzed. One way to plot the data is shown in Figure 6.9. Statistical analysis of the data is often required to interpret the data and extrapolate them to larger sample sizes that correspond to a product that is produced in high volumes. This is discussed in more detail in Nelson (2004).

10.4.5 Review Data and Select Materials

After the materials evaluation and process development are complete, a design team can select the materials that optimize the performance, reliability, and cost of the product. As mentioned several times earlier, the cost includes the cost per unit of the material and costs of poor quality.

10.4.6 Write Specifications for the Materials

The material features, properties, and defects information that should be in the specifications for subassemblies, components, and primary input materials was discussed in Chapter 7.

10.4.7 **Other Situations for Selection of Materials**

In addition to selecting materials during product development, the need to do so arises during the following situations for a product already in production:

Cost reduction. This involves replacing one material or process with a different material or process that reduces total costs. Total cost includes the cost per unit item, manufacturing costs, and costs of poor quality (e.g., scrapped product and product failures).

Product reliability improvement. This involves replacing a material that does not enable the product to meet its reliability requirements with a material that does enable the product to meet the requirements.

Manufacturing process yield improvement. This involves replacing a material (or process) that contributes to a lower-than-desired process yield with a material or process that enables greater process yield.

Replacing an obsolete, bottleneck, or banned material. A material may become obsolete because its manufacturer stops producing it. Replacing a bottleneck material addresses a sourcing issue related to the need to replace a material that is difficult to obtain or one with which few suppliers have experience. A material can be banned by government regulation for health, safety, or environmental reasons. In all of these cases, it is up to the design team to find a replacement material that has properties and costs that satisfy the design requirements of the product element. If there are no other options, then the design team will have to either redesign the product element or be willing to modify the product elements' design requirements. The latter option may require modifying the design requirements for the entire product.

10.5 MANUFACTURING PROCESS DEVELOPMENT

Manufacturing process development involves determining the process inputs required to develop a capable method for producing process output that consistently meets its design requirements. The process output can be a component, subassembly, or assembly made by a Type I, Type II, or Type III company. The process inputs are determined using experiments designed to reveal the relationship between the process inputs and the material features, material properties, performance, and reliability of the product elements being formed. The experiments include evaluating the effects of using different input materials, the variations in the input materials, different types of process equipment, and different process parameter set points.

Process development is coupled to materials selection. It may be determined that out of three input materials, one is easier to use than the others. The material may enable the process to run at a faster rate or result in fewer defects in the product element. These are important factors to consider when selecting a material.

> ### Manufacturing Process Development Goals
>
> **General Goals**
> - Develop manufacturing processes that consistently produce an output that meets its design requirements.
> - Develop manufacturing processes that operate at the lowest possible cost.
>
> **Materials Engineering Goals**
> - Understand the relationship between the manufacturing process inputs and the material features, material properties, performance, and reliability of the output's product elements.
> - Understand the effects of variation of the manufacturing process inputs on the variation of the process outputs.

10.5.1 Process Development Goals

There are two general goals for process development. The first is to establish manufacturing processes that consistently produce an output that meets its design requirements—that is, manufacturing processes that have good yields. This is achieved by determining the optimum process inputs and learning how to control their variations. The second goal is to develop manufacturing processes that operate at the lowest possible cost. This is achieved by using processes that are as simple as possible, using low-cost input materials, and accomplishing the first goal.

From the materials engineering perspective there are two goals. The first is to understand the relationship among the performance, reliability, materials properties, and materials features of the components, joints, or in-process structures being formed and the input materials, process equipment, and process parameters used to form them. The second goal is to understand the effects of manufacturing process input variation on process output variation.

A description of the relationship between the process inputs' effect on the output can be quantitative or qualitative. Graphs and equations provide quantitative descriptions and micrographs provide qualitative descriptions. An example of a graph is presented in Figure 10.5, which shows the relationships between three hypothetical input variables—A, B, and C—and an attribute of the output. As Input A increases the output attribute increases. As Input B increases the output attribute decreases. As Input C increases the output attribute increases, passes through a maximum, and decreases.

Micrographs showing the changes in the microstructure of a material for different values of inputs is another method for displaying the effects of the process inputs on the output. For example, the strength of a braze joint depends on the

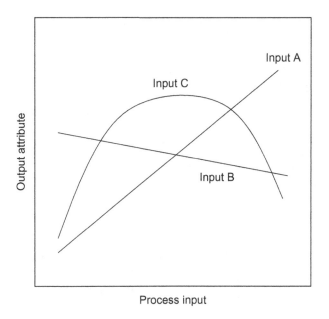

FIGURE 10.5

Graph showing the relationship between a process output attribute and three process input variables.

extent of the metallurgical reaction between the braze alloy and the materials being joined. Understanding the relationships among the brazing temperature, brazing time, joint strength, and joint microstructure enables the design team to make better-informed changes to the braze time or temperature to obtain the microstructure that provides the desired joint strength.

This information, which relates the inputs to the outputs, has several uses. First, it can be used to fine-tune a manufacturing process by enabling the selection of the optimum set of process inputs to achieve the desired output with the maximum process capability. The second use is for problem solving. Once a process has been implemented and the item is in production, the information obtained during process development allows production yield problems to be resolved more quickly. A decrease in the production yield of an item can be related to problems with the quality of input materials, process equipment malfunctions, or operator error. These problems will be manifested as material features, properties, or defects that do not meet requirements for one of the product elements being formed.

The process development information can be used to determine the specific changes of the inputs that led to the changes in the output. Without this information it is difficult to determine which input has changed. Finally, the input–output information for the existing process can be used as a starting point for making process improvements or developing new processes based on the existing process.

10.5.2 Process Development Procedure

The following process development steps are shown in the flowchart that is in Figure 10.6.

1. Identify the manufacturing process steps.
2. Identify the process input and output variables.
3. Determine the values of the process input variables.
4. Write specifications for the input materials, process equipment, process steps, and output analysis methods.

The elements of a specific process effort depends on whether the manufacturing process is new to an organization or already exists. For a new process, all of the steps will be followed. For a preexisting manufacturing process, some or all of these steps may be unnecessary. The items that must be addressed depend on the similarity between the new product and other products being made using the current process.

10.5.3 Identify the Manufacturing Process Steps

A manufacturing team must identify each step of a manufacturing process, starting from setting up all of the input materials to prepare them to go through the

Detail design—manufacturing process development

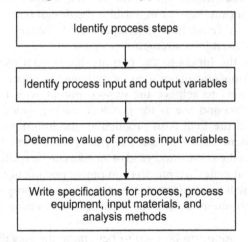

FIGURE 10.6

Flowchart for manufacturing process development.

product line to evaluating the output at the end. In between, the process consists of one or more steps to form the product elements. For example, brazing two components together might involve the following steps: (1) applying flux to the materials to clean the surfaces, (2) applying braze compound to the components, (3) placing the components together, (4) placing the components in an oven, (5) removing the components from the oven, and (6) evaluating the joint strength of a sample.

As each step of the process is defined, the manufacturing team should identify the problems that can occur with the operators, equipment, and materials that will adversely affect the output. Then steps should be put in place to prevent the problems from occurring or to allow for their easy detection. Developing a process failure mode and effects analysis (PFMEA) will help identify potential sources of problems and how to prevent or detect them. Failure modes and effects analysis (FMEA) was discussed in Chapter 7 as it applies to developing specifications. The same principles can be applied to any process to determine failure modes, effects of the failure, the cause of the failures, and methods to prevent and detect the failures.

10.5.4 Identify the Process Input and Output Variables

As discussed in Chapter 5, the input variables for a manufacturing process are the input materials, process equipment, process conditions, operator skill, manufacturing environment, and data feedback from evaluations of the output. The output variables include the performance attributes, reliability attributes, materials features and properties, and defects for the item being produced and its product elements.

The manufacturing team must identify the input variables that need to be controlled and the output variables that should be evaluated. The significant variables depend on the design requirements of the output and the specific processes used to make it. For the same type of process, it will be necessary to control more input variables for an item that must be made to close tolerances compared to a more forgiving item. Also, the number of evaluations of the more demanding item will be greater than that for the more forgiving item.

If a product line does not already exist, then the manufacturing team will have to identify the specific process and output evaluation equipment required. The specific process equipment will depend on the controls and stability required to develop a capable process. The specific output evaluation equipment will depend on the required measurement capability to evaluate the process output.

10.5.5 Determine the Values of the Process Input Variables

The values of the input variables that need to be determined are the specific input materials, process equipment, process parameter set points, and output evaluation equipment. After selecting the process and output evaluation equipment, the materials and process parameter set points are determined by performing experiments to determine the effects of the materials and process parameters on the output variables of interest. In many cases, manufacturing teams use design of experiments (DOE) to help set up the experiment. DOE is a method for designing experiments so that they are efficient in terms of time, money, and effort. It is especially helpful when there are more than just a few input variables for which values need to be selected. Information about DOE is available in Mason (2003), Montgomery (2004), and Ross (1995).

Verify Materials

Before performing any process development experiments, a manufacturing team should analyze the input materials to verify that they are what they are supposed to be. This prevents the team from spending money and time conducting experiments that may be unsuccessful because the materials do not meet the requirements, and it prevents the team from making decisions based on erroneous information. It is not uncommon for manufacturing teams to go into production and learn that the materials are different than those used during process development. Obviously, this situation is undesirable.

Evaluate Variations of Input Materials Properties

As mentioned earlier, the properties of a material can vary from supplier to supplier. Therefore, for a particular type of material, it is necessary to evaluate the effects of materials from various suppliers on the output variables. The supplier whose material enables the widest process window should be selected. The information in Section 10.4.2 about the variation of material features, properties, and defects applies to this discussion.

10.5.6 Write Manufacturing Process, Input Materials, and Output Evaluation Methods Specifications

An output of the process development process is a set of specifications for the manufacturing process, input materials, and output evaluation methods. The information to include in each of these specifications was discussed in Chapter 7.

10.6 COMPLETE SUBASSEMBLY AND COMPONENT SPECIFICATIONS

Once the designs for all of the components and subassemblies are complete, a design team can finalize the component and subassembly specifications. The information to include in the specifications was discussed in Chapter 7. Specifications should also be written for off-the-shelf components and subassemblies approved for use.

Product Verification Testing Goals

General Goals
- Determine whether a product satisfies its reliability requirements.
- Determine the root cause of test sample failures.

Materials Engineering Goals
- Determine whether the materials that make up a product satisfy the product's reliability requirements.
- Determine the root cause of test sample failures.

10.7 PRODUCT VERIFICATION TESTING

As discussed in Chapter 6, product verification testing involves exposing samples of a product to conditions that simulate its use conditions and then determining whether the test samples still meet the performance requirements. If a sample fails a test, then the design team must identify and eliminate the root cause of the failure. Addressing the root cause may involve modification of the design or the manufacturing processes. The tests are usually repeated after the root cause of the failure is identified and fixed. Ideally, a product will pass each set of tests the first time, eliminating the need for product redesigns or manufacturing process modifications. However, this is not always realistic, especially for the development of new platforms or fundamentally new products.

Verification tests are performed at various points of the product development process. During the early stages, the tests are used to uncover design flaws. At the later stages, they are used to verify that design and manufacturing processes are adequate.

The general goals of product verification testing are (1) to determine whether a product satisfies its reliability requirements and (2) to determine the root cause for test sample failures. The materials engineering goals are (1) to determine whether the materials enable the product to satisfy its reliability requirements and (2) to determine the root cause for test sample failures.

10.7.1 Designing Product Verification Tests

For a Type I company, selecting the appropriate verification test conditions is critical to obtain valid data. Because it is desirable to obtain the data in as short a time as possible the tests often involve accelerated exposure of the test samples to the use conditions, as discussed in Chapter 6.

When designing product verification tests, it is important that there be a direct correlation between the physical behavior of the materials exposed to the test conditions and those exposed to the normal use conditions. When using accelerated test conditions they should not be so aggressive as to cause components, joints, or in-process structures within the test samples to degrade and fail in a manner unlike that caused by exposure to the product use conditions. Conversely, if the test conditions are not aggressive enough, then the samples may pass the tests but fail to meet the reliability requirements given in the product specification. Both situations were discussed in Chapter 6.

The verification tests for a Type I product should be specified when the product reliability requirements are established during the product concept development phase. The test conditions, duration of test, types of samples, number of samples, and test procedure should be outlined early in the product development process. This helps design teams develop realistic product development schedules. Many products (e.g., medical devices or safety devices) must include testing specified by industry standards or government regulations.

Type II companies may have to perform product verification tests as directed in the subassembly or component specification provided by the client company. Type II companies do not have to develop the tests. There can be a negative side to this, however. Sometimes, a Type II company believes that the specified verification tests are too aggressive compared to the actual use conditions. In many cases, the Type II company may discover this long after its proposal to design and build a component or subassembly has been accepted. Then, the design team from the Type II company must overdesign its product in order to pass the verification tests, increasing the cost to make the product. Unfortunately, there may be little that the design team can do. This is another reason why a materials engineering perspective is needed when a Type II company design team writes a proposal to design and manufacture a component or subassembly according to the requirements from another company.

10.7.2 Performing Verification Tests

When performing product verification tests, a design team must verify that the composition, microscopic features, and properties of the materials to be tested meet the requirements of those to be used in the intended product. Also, the manner in which the tests are performed should be monitored to verify that the

---

Root Cause Analysis Goals

General Goals
- Identify the root cause of a failure or problem.
- Put systems in place to prevent the root cause from recurring or detect it when it does occur.

Materials Engineering Goals
- Identify the root cause of a failure or problem and fix the problem.
- Put systems in place to prevent the root cause from recurring or detect it when it does occur.

test conditions are as they should be. These verifications prevent false failures or false passes when performing tests with materials that do not meet the requirements, and was discussed in detail in Chapter 6.

10.8 **ROOT CAUSE ANALYSIS**

Root cause analysis is concerned with identifying the origin of failed product verification test samples, product failures during customer use, poor manufacturing yields and poor supplier quality. The root cause is the fundamental reason for a failure or defect and is associated with one or more product elements that do not perform as required. For example, the steel shaft of a motor may break because of fatigue caused by the presence of excessive nonmetallic inclusions within the steel, which resulted from an improper steel-making processes caused by the malfunction of the steel-making equipment or operator error. In this example, the equipment malfunction or operator error is the root cause of the shaft failure. The inclusions in the steel are the symptoms of the root cause.

Failures arise from various root causes including poor design, inferior materials, suboptimum or poorly controlled manufacturing processes, customer misuse, and wearout. Wearout is when a material is used beyond its expected life span. The root cause of product verification test failures can also be poorly controlled test conditions. When a failure occurs, design and manufacturing teams identify the root cause of the problem and put systems in place to prevent it from recurring or to detect it when it does occur.

In many cases, a failure is related to a problem with the materials that constitute one or more product elements within the failed product. Therefore, root cause analysis often includes analysis of the materials that make up the product elements associated with the failure. The analysis can yield the following information: (1) whether there was degradation and, if so, the mode of degradation; (2) the product element's failure mode; and (3) the features, properties, and defects of the materials. Only after this information is obtained is it possible to identify the root cause of the failure.

At the beginning of the root cause failure analysis, engineers (1) identify the product function that was lost or the malfunction that occurred and (2) determine the test or use conditions when the failure occurred. Next, the engineers try to identify the particular subassembly or product element associated with the failure. If the failure originated within a particular subassembly, then the engineers identify the failed product elements within the subassembly. If the failed subassembly was manufactured by another company, then the subassembly may be sent to that company for further analysis.

After identifying the product elements associated with the failure, the next step is to identify their failure mode. Sometimes engineering teams spend a great deal of time trying to develop hypotheses about the failure mode and its root cause before obtaining the information from a materials failure analysis. This effort is typically not productive because using intuition or past experience to guess about the cause of a material failure mode is almost impossible.

Certain information about the failed subassembly or component should be provided to help the analyst understand the failure mode and potentially assign the root cause of the failure. This information includes the following:

- Component and subassembly drawings
- Description of how components were made
- Technical data sheets for off-the-shelf components, joining, or coating materials involved in the failure
- The normal use conditions of the subassembly or component
- The conditions under which the subassembly or component failed

In addition, it is often helpful to provide samples of subassemblies or components known to be good as well as samples from the same production lot as the failed samples. Materials characterization data collected during the materials and product verification testing are also helpful to establish whether a substantial change occurred in the materials or processes used between the time of the product verification testing and the field failure.

When providing the failed samples and information to the materials analyst, the engineering team should be careful about framing the question to be answered (i.e., why did the sample fail?). Engineering teams sometimes make the mistake of trying to direct the analyses to perform or the strategy for the failure analysis. This practice is best avoided because it risks sending the analyst down the wrong path or getting only partial information that does not allow the cause of the failure to be identified.

If the analyst takes the wrong path, then time and money is spent gathering information that does not help solve the problem, or what is thought to be the root cause of the failure may be identified only to discover much later that it is not the right one. If the engineering team gets only partial information, the team may think it has hit a dead end or may realize that more analyses must be done. In the second case, the team may eventually obtain all of the required materials information about the failed sample, but only after spending more time than should have been necessary. It is usually best to just review the failure and materials information with the analyst and ask the person to identify the failure mode.

Once the samples have been analyzed and the product element's failure mode has been identified, the next step is to determine the root cause of the failure. Sometimes the cause is readily apparent; other times, more investigation into the design, manufacturing processes, or suppliers is required.

More information about materials failure analysis is available in *ASM Handbook*, Volume 11 (2002) and Colangelo et al. (1987).

REFERENCES

ASM Handbook, Volume 11: Failure Analysis and Prevention, ASM International, 2002.

Colangelo, V. J., and F. A. Heiser, *Analysis of Metallurgical Failures*, 2nd edition, Wiley-Interscience, 1987.

Mason, R. L., R. F. Gunst, and J. L. Hess, *Statistical Design and Analysis of Experiments, with Applications to Engineering and Science*, 2nd edition, Wiley-Interscience, 2003.

Montgomery, D. C., *Design and Analysis of Experiments*, 6th edition, Wiley, 2004.

Nelson, W. B., *Accelerated Testing: Statistical Models, Test Plans, and Data Analysis*, Wiley, 2004.

Ross, P. J., *Taguchi Techniques for Quality Engineering*, 2nd edition, McGraw-Hill Professional, 1995.

Production

11.1 INTRODUCTION

Once a product goes into production, design and manufacturing teams work on the following tasks:

- Increasing the manufacturing yield
- Reducing the costs to manufacture the product

Both are part of continuous improvement efforts, which are necessary for the product and company to remain competitive. This chapter discusses the application of the materials engineering perspective to each of these tasks.

11.2 IMPROVE MANUFACTURING YIELD

The holy grail of manufacturing organizations is when all of the manufacturing process output satisfies its design requirements (i.e., 100% yields). Unfortunately, this rarely occurs, even after the most thorough process development. When a product first goes into production (i.e., the product launch), the yields will probably be lower than desired, and the output that does not meet its design requirements is either fixed or thrown away (i.e., scrapped).

Improve Manufacturing Yield Goals

General Goals
- Eliminate systemic causes of chronically low manufacturing yields.
- Eliminate the root cause of sudden drops in manufacturing yields.

Materials Engineering Goals
- Improve the understanding of the correlation between the process input variables and the process output variables.
- Improve the process and input materials' specifications.

Low yields are due to nonoptimized process inputs, excessive variation of one or more process inputs, or suboptimum design of the process output. The number of samples that can be made and evaluated during product and process development is usually just not large enough to uncover the cause of every output variation. Therefore, once a manufacturing line starts making the product in large quantities, it is not unreasonable to expect more output variation than observed during process development. Also, the increased amount of output produced may expose problems with the product's design or the process inputs.

Manufacturing and design teams work to improve manufacturing yields. Once the yields reach acceptable levels, there are sometimes unpredictable changes in a process input that lead to a sudden decrease in the yield.

Addressing low manufacturing yields often requires a materials engineering perspective because in many cases the root causes of yield problems relate to the input materials or the influence of the other manufacturing process inputs on the materials features and materials properties of the product elements that make up the process output.

11.2.1 Effects of Low Yields

Low yields result in reduced profits through two mechanisms. First, sales will drop if there are not enough samples of the product available to go on the market. This occurs if the total number of samples that the production line can manufacture is limited so that it is not possible to make extra product samples to compensate for the scrapped output. Second, there are costs associated with making a defective product that will be thrown away. These costs are related to the following factors:

- The costs to purchase the materials, components, or subassemblies used to make the defective product.
- The excess production capacity required to generate output that can be sold plus the output that must be scrapped. That is, more samples than necessary must be produced because a fraction of the total number of samples produced will be defective.

Increasing the yields may enable the use of less manufacturing equipment and fewer workers, while reducing the number of materials, components, or subassemblies that must be purchased to make the same amount of sellable product.

11.2.2 Types of Yield Problems

There are two types of manufacturing yield problems: (1) chronically low yields and (2) sudden yield drop. Chronically low yields occur when, over time, the yields are consistently lower than those necessary to make the desired profit. A sudden drop in yield is caused by an abrupt change in a process input so that it

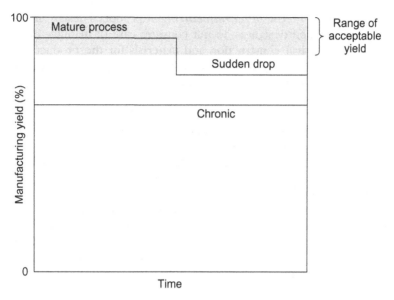

FIGURE 11.1

Manufacturing yield versus time.

does not satisfy its requirements, resulting in a sudden change in the value of one or more output variables so that they do not satisfy their requirements. Figure 11.1 shows both types of yield problems in a graph of manufacturing yield versus time.

Chronically low yields are due to one or more of the following causes:

Using suboptimum process inputs. This is a result of poor materials selection and poor process development. Examples of problems resulting in chronically low yields are listed here for the different categories of process inputs:

- *Input materials:* Materials suboptimum for ease of manufacturing, excessive variation of the material features and properties
- *Process equipment:* Excessive instability
- *Output measurement equipment:* Inaccurate, excessive instability
- *Process equipment fixtures:* Suboptimum design (i.e., suboptimum materials, suboptimum physical construction)
- *Process parameter set points:* Not optimized
- *Operator:* Poor training

Consistently excessive process input variations (i.e., common cause variations). This occurs with input materials if the supplier has poor quality control, process equipment that is incapable of maintaining the process parameters close to their set points, and a poorly controlled manufacturing environment.

Not designing the product for ease of manufacturing. A product's design affects manufacturing yields when designers do not consider ease of manufacturing when selecting the physical construction and materials for the product elements. An example is designing two components without considering how they have to fit together so that they can be easily joined.

Chronically low yields are typically associated with production ramp-up, which is the period of time following the introduction of a new product or manufacturing process into a production facility. During this period, the manufacturing team scales up the amount of production output from the small batches produced during product and process development to the larger quantities necessary to meet market demands. This involves increasing the production line's capacity and yield. Except for the simplest manufacturing processes, it is unrealistic to expect a flawless ramp-up, especially for a brand new production line.

Sudden drops in manufacturing yields occur when there is a special cause variation of a process input. Examples of special cause variations are changes in the properties of input materials as a result of problems with a supplier's production line, an equipment malfunction, incorrect setting of a process set point because of operator or automation error, or seasonal changes in the manufacturing environment.

11.2.3 Goals for Improving Manufacturing Yields

The general goals for improving manufacturing yields are (1) to eliminate systemic causes of chronically low yields and (2) to eliminate the root cause of sudden drops in yields. From the materials engineering perspective, the goals are (1) to improve the understanding of the correlation between the process input variables and the process output variables and (2) to add more detail to the process and materials specifications.

11.2.4 Improving Chronically Low Yields

To improve chronically low yields, manufacturing teams engage in a continuous improvement process during which they gain knowledge through experiential learning and controlled experiments using the production line as a laboratory. Experiential learning is knowledge gained while making improvised corrections to design and process problems. The trouble with this type of learning is that knowledge is gained at the expense of continued poor production yields. The problem with performing controlled experiments is that they use production capacity, creating a trade-off between regular production for revenue and experimentation for learning. Extended and unprofitable periods of chronically low yields can exist when managers assume that learning will occur automatically through experience and therefore underinvest in deliberate learning through experimentation. This often occurs during production ramp-up.

For a mature manufacturing process, manufacturing yields can be close to 100%. However, except for the simplest of products, obtaining 100% yields is difficult because identifying and eliminating all sources of process input variation is challenging and perhaps impossible. As a practical matter, it becomes more difficult and expensive to control all sources of common cause variation as the yield approaches 100%. Therefore, for a particular manufacturing process, yields of less than 100% are usually acceptable as long as they are within a certain range. The decision regarding whether or not to invest resources in improving manufacturing yields depends on the estimated costs to identify and implement the design or process modifications necessary to make the improvements.

Improving chronically low manufacturing yields involves many of the concepts applied to process development and root cause analysis, which were discussed in Chapter 10. For output that is an assembly or subassembly, manufacturing teams must identify the product elements associated with the output's problems and then determine the root causes of the problems. This involves the following steps:

1. Identify the product elements associated with the problem. If the assembly or subassembly contains subassemblies, then the problem product elements can either be joints between the lower-level subassemblies or a product element within the lower-level subassemblies.

2. Evaluate the performance, shape, dimensions, material features, material properties, and defects of the product elements to determine whether they satisfy their design requirements.

3. If the analyses indicate that all of a product element's design requirements are satisfied, then either the problem is not associated with the product element or the product element has been poorly designed.

4. If the analyses show that one or more of a product element's design requirements are unmet, then the manufacturing team must identify the root cause for the problem with the product element.

5. Determine the process input modification or controls required to eliminate the root cause of the problem with the product element. Modification is required if any of the input variables are suboptimum. Identifying the input modification may require experimentation to determine the input variable to modify and the required change. Greater control is required if there is too much variation of the input variables or if it is possible to use inputs that are not as specified (e.g., an operator using the wrong process parameter set point).

If the product line's output is a component, the manufacturing teams must identify the materials within the component that are associated with the component's problems and then determine the root cause for any problems with the materials. This involves steps 4 and 5 from the root cause analysis process for an assembly or subassembly.

The data from process development help improvement efforts by serving as a benchmark against which the production data can be compared. Differences between the two sets of data and knowledge of the process inputs may indicate the process input variables that need to be investigated and adjusted in order to match those used during process development.

11.2.5 Fixing Sudden Yield Drops

To determine the root cause of a sudden drop in yields, the team will perform a root cause analysis similar to that used for chronically low yields. However, special cause variation problems can sometime be more elusive. Unlike chronically low yields, which continue until the root cause is identified and remedied, special cause variation problems can sometimes appear and disappear without any action from an engineering team, continue until the root cause is remedied, or be intermittent.

The process for identifying the root cause of a sudden yield drop is the same as that for chronically low yields. However, the actions once the root cause is found are different. Instead of modifying a process input, an engineering team must put systems in place to prevent the special cause variation from recurring. For example, addressing an equipment problem might involve a periodic maintenance check, a problem with a supplier's product might involve added inspections by the supplier before shipping it, and a problem associated with wrong actions by an operator might involve additional training.

11.2.6 Update Specifications

After the root cause for a yield problem has been identified and addressed, the engineering team should update the relevant specifications to include the changes to the process inputs or the control systems put in place.

Cost Reduction Goals

General Goal
- Reduce the cost to manufacture a product without compromising the product's performance, reliability, or manufacturing yields.

Materials Engineering Goals
- Understand the effects of process, material, and supplier changes on the material features, properties, performance, and reliability of the product elements associated with the change.
- Understand the effects of the proposed changes on the manufacturing process capability and yields.

11.3 **COST REDUCTION**

After a product goes into production, design and manufacturing teams look for ways to reduce the cost to manufacture it. The main contributors to the costs to produce a product are (1) the cost of the materials, components, and subassemblies that comprise the product and (2) manufacturing costs such as the cost of the manufacturing equipment and space, labor costs, and costs of poor quality.

The general goal of a cost reduction effort is to lower the cost to make a product without compromising its performance, reliability, or manufacturing yield. For example, a weld joint made between components purchased from a new, lower-cost supplier should be formed as easily as a weld joint made between components from the previous supplier, and the weld joints should have the same properties as before. There are always risks that the change will cause problems, and any associated costs could counter the expected savings. It is up to design and manufacturing teams to determine the risks and to develop an evaluation plan that addresses them.

From the materials engineering perspective, there are two goals. The first is to understand the effects of process, material, and supplier changes on the material features, properties, performance, and reliability of the product elements associated with the change. This knowledge helps teams to understand and mitigate the risks associated with the proposed changes. The second goal is to understand the effects of the proposed changes on the manufacturing process capability and yields.

11.3.1 **Methods to Reduce Costs**

Cost reductions can be achieved through several approaches, which fall into the following general categories:

1. Modifying designs
2. Changing suppliers of materials, components, and subassemblies
3. Modifying manufacturing processes
4. Increasing automation
5. Using lower-cost manufacturing facilities

Each of these categories is discussed next.

Modifying designs. There are several approaches to cost reduction by design modification. From the mechanical and electrical engineering perspective, design modification involves redesigning a subassembly or assembly so that one or more components can be eliminated or replaced by less expensive or fewer components. From the materials engineering perspective, design modification includes a few more options, as follows:

- Changing to less expensive materials that still satisfy a product element's design requirements. This is possible if the team determines that the materials currently being used offer better performance and reliability than required or if a lower-cost material similar to the one in use becomes available.
- Redesigning the physical construction of a component so that less material is required to make the component. This can be significant if the material is expensive. The same redesign strategy can be applied to use a material that is more expensive than the current material, but less of it is required, reducing the overall cost. For example, using a stronger material makes it possible to reduce the diameter of a motor shaft.
- Using materials that increase manufacturing yields or the amount of output created per unit time. For example, for a particular production line it may be easier to form good ultrasonic weld joints between plastic components made of different types of plastic compared to the current material. Using the alternate plastic will make the process more capable, resulting in higher yields. The change may also result in a decrease of the time needed to weld the components, which will enable more output to be produced in a given period of time compared to the output for the current material.
- For a platform of similar products, using common materials for components that are similar from product to product. The increase in usage for a particular material often results in volume discounts from the material supplier.

Changing suppliers of materials, components, and subassemblies. There are two ways to reduce costs by changing suppliers. The first method is to select a supplier that can provide a lower-cost material, component, or subassembly that is similar to that currently being purchased. The second approach is to select a new supplier that can provide a material, component, or subassembly that is of better quality and has less variation than that offered by the current supplier. This leads to a more capable manufacturing process that has higher yields, which corresponds to lower costs to produce the same amount of sellable product.

Modifying manufacturing processes. This involves modifications to the process equipment or fixtures, process parameters, or process steps. Any of these changes is made to improve manufacturing yields, increase production line throughput, or increase production line capacity. Increasing yields lowers costs by reducing the amount of waste associated with purchasing material, components, and subassemblies that end up in defective process output that must be thrown away. Increasing production capacity enables the use of fewer operators or shifts to generate the same amount of output that was produced before the process modification, thus reducing labor costs. Alternatively, increased manufacturing capacity allows for more of the product to be produced with the same number of operators and shifts, and the same amount of equipment. This is important for a product in high demand because it reduces the cost of labor per item produced and increases the number of items available for sale.

Increasing automation. This option reduces labor costs. The automation can be for equipment that handles the input materials, components, or subassemblies for the product. Automation can also be used to control the process equipment and setting and to maintain the process parameters. In addition to reducing the need for manual labor, increased automation also reduces common cause variation and increases yields. The manner in which processes are performed is the same for every sample passing through an automated production line. To derive this benefit, there must be good control over the variation of the input materials. If there is not, then an automated production line enables a great deal of defective output to be produced in a short period of time.

Using lower-cost manufacturing facilities. Many companies move their production lines to countries where the cost of labor is cheaper than it is in the original country.

Sometimes a change includes benefits that are ancillary to the cost reduction effort. For example, changing to a less expensive material used to make a component may also reduce the time required to form a joint to the component, if the new material is easier to weld, solder, braze, or adhere to. The secondary benefit is increased manufacturing capacity and throughput.

11.3.2 Evaluating the Effects of the Change

There may be risks associated with a proposed cost reduction effort. These risks relate to implementing a change that results in a less capable process with reduced yields or reduction in the reliability of the process output. These risks can be mitigated by evaluating the effects of the proposed changes on the material features, material properties, shape, dimensions, performance, and reliability of the affected product elements and on the performance and reliability of the product.

It may seem obvious that changes should be tested and verified. However, many decisions have been made "intuitively," with disastrous results. In one instance, a decision was made to modify a component cleaning process based on an intuitive assumption. Two months later, it was discovered that the components did not have the proper electrical characteristics. Further analysis showed that the change in the cleaning process altered the surface chemistry of the component, which affected the process step that followed the cleaning step. The cost of this mistake was approximately $100,000 worth of defective components that could not be sold.

Evaluating the effects of a change requires a procedure that is similar to that used for selecting a material and process development. The evaluations that follow should be performed.

Evaluate the feasibility of the change. This assessment involves determining the effects of the change by analyzing the material features, material properties, defects, performance, and reliability of the affected product elements. This appraisal also involves materials reliability testing when considering a new material, process modification, or new supplier. Consideration of a new supplier should include an evaluation of its manufacturing expertise and competence.

Verify that the change does not have a negative impact on the product's performance and reliability. This assessment involves evaluating the effects of the change on the performance and reliability of the product. Product verification tests must be performed for many products after a design, material, or manufacturing process modification and when changing suppliers.

Evaluate the effects of the change on manufacturing process capability and yield. This assessment involves verifying that the change has no adverse impact on the manufacturing processes. Process development may be necessary to determine the changes to the process and inputs necessary to implement the proposed change.

11.3.3 Costs of the Change

Any cost reduction effort requires an investment of resources to evaluate the effects of the proposed changes. There are costs associated with making test samples and performing the evaluations. These costs must be considered before starting a cost reduction project in order to determine the overall cost benefit of the change. Then a cost reduction team must determine whether the projected savings justify the costs associated with evaluating the impacts of the proposed change. The evaluation costs are directly proportional to the risk associated with the change. As the risk increases, the number of evaluations required to verify that the change will not cause problems increases as well.

Materials Engineering Strategies for the Product Realization Process

12

12.1 INTRODUCTION

Many pressures are associated with developing and manufacturing a product. It can be challenging to meet all of the design requirements and the project schedule and still have good manufacturing yields. Additionally, keeping costs low is always a concern. Many of the problems encountered during product development and manufacturing can be avoided by using strategies based on the materials engineering perspective:

1. Start with materials that offer a high probability of success.
2. Do not consider the entire world of materials, components, and subassemblies for use in a product.
3. Work out all details of a strategic custom component or subassembly before using a low-cost supplier.
4. Consolidate materials within and across product platforms.
5. Develop design guidelines.
6. Include a budget for materials engineering support.

Some of these strategies may seem restrictive. However, they force engineering teams to be more disciplined and to seek more creative options.

12.2 START WITH MATERIALS THAT OFFER A HIGH PROBABILITY OF SUCCESS

One way to reduce the pressures and risks associated with developing a new product is to use materials that offer a high probability of satisfying the product element design requirements but may not be the lowest-cost options. This strategy improves the probability of meeting product development deadlines and developing capable, high-yield manufacturing processes. Lower-cost materials can be substituted after it has been verified that the product has met all of its performance and reliability requirements. In some cases, it may be possible to make the substitution before the product launch. In other cases, the substitution will have to be made after the product has been in production for some time.

Trying to use specific materials can constrain the options for the mechanical or electrical design. For example, specific materials may require a certain physical construction for a component, or they may require the use of off-the-shelf components that consist of a specific metal that enables the formation of good solder joints to the component. Or the material may need to be of a certain thickness for a component to have the necessary strength. There may be some resistance to constraining the mechanical or electrical design because it may limit design flexibility, from the mechanical and electrical engineering perspectives.

This strategy has a few benefits. First, it eliminates the need for materials reliability testing. Second, if the production line has experience with these materials, the need for process development is minimized. Both of these benefits allow engineering resources to be devoted to other tasks and decisions. Finally, this strategy improves the probability that the product will pass any particular round of product verification tests the first time they are run.

Realistically, this approach may be acceptable for some portions of a design, but not all. For some product elements, there may be no option other than considering higher-risk materials.

12.3 DO NOT CONSIDER EVERY MATERIAL, COMPONENT, AND SUBASSEMBLY IN THE WORLD AS OPTIONS FOR A PRODUCT

Sometimes, a design engineer will identify a material or an off-the-shelf component or subassembly to use in a product, but the engineer and the organization has no experience with the material or the secondary input materials within the component or subassembly. This poses risks with regard to the materials meeting performance, reliability, and manufacturing requirements. Even if the materials are common, it is still necessary to verify that they will consistently have the required features and properties. As discussed in previous chapters, it is important to remember that even common materials can have different properties depending on how they were made and who made them.

A different strategy is to consistently use only approved materials, off-the-shelf components, and off-the-shelf subassemblies within a common product platform. This may seem restrictive; however, it will force design teams to be more creative. Not using this strategy burdens other groups within the organization, such as manufacturing and quality engineers, with having to determine how to make the product using suboptimum materials. This may lead to product delays and low manufacturing yields.

12.4 **WORK OUT ALL CUSTOM COMPONENT OR SUBASSEMBLY DETAILS BEFORE USING A LOW-COST SUPPLIER**

When selecting suppliers of strategic materials, components, and subassemblies for a new platform of products or a fundamentally new product, technical and manufacturing expertise should be given a higher consideration than cost. After the design and manufacturing details have been worked out, then a design team can consider other suppliers that can provide their products at a lower cost. This strategy improves the chances of developing the product on time and with fewer problems. From the materials engineering perspective, this approach improves the likelihood that the materials and manufacturing processes used by a supplier will enable the component or subassembly to meet its design requirements.

For a new product, there are typically many design and manufacturing unknowns for strategic components and subassemblies. These unknowns pose risks to the development schedule and to the likelihood of generating a product that satisfies all of its requirements. Using a low-cost supplier that does not have a proven ability to mitigate the risks for a specific component or subassembly adds to the overall risks and may end up causing the total costs to increase because of problems encountered during product development and after the product is launched. When this occurs, the supplier may no longer be low cost.

12.5 **DEVELOP DESIGN GUIDELINES**

Within a product platform and sometimes across platforms, an approved set of materials may be used from product to product for specific product elements. However, differences in the design requirements for the various products may require some flexibility in (1) the physical construction of a product element using a specific material or (2) the options of materials that can be used for a common product element. An example of the first situation is allowing for different shaft diameters for motors with different powers but made from the same material. An example of the second situation is a choice of two different silicone materials for use in a windshield wiper insert. Selection of one material over another may depend on the desired cost of the insert or on the reliability requirements.

For either situation, the design process can be sped up if there are design guidelines that indicate the materials that have been approved for a product element, the corresponding approved physical construction of the product element, and the applicable design requirements. Based on these requirements, a set of guidelines enables a design team to quickly select the materials and physical construction for a product element. Design guidelines are based on knowledge of and experience with the performance and reliability of the materials and any of the other design requirements.

Table 12.1 Hypothetical Electrical Wire Design Guidelines for Maximum Temperature Rise of 30°C

Material	Minimum Diameter (mm)	Maximum Current (Amps)	Operation Environment Temperature (°C)
Alloy 1	2	40	35
Alloy 1	4	80	35
Alloy 1	3	40	85
Alloy 1	6	80	85
Alloy 2	1.5	40	35
Alloy 2	3	80	35
Alloy 2	1	40	85
Alloy 2	2	80	85

The information that may be incorporated into a design guideline can include the following:

- The design requirements that the product element must satisfy. This includes performance, reliability, cost, manufacturing, and the other categories.
- A list of approved materials.
- The product element's physical construction.
- Descriptions of the composition, microscopic structure, surface condition, and defect requirements of each material.

A design guideline can be in the form of a table, graph, or equation. An example of a design guideline table is shown in Table 12.1. This table, which contains hypothetical data, indicates the required wire minimum diameter to ensure that the temperature of the wire increases no more than 30°C above the operating environment. A potential application of this information is for the design of power electronics containing wire. Figure 12.1 is a graph of hypothetical data for wire diameter versus current. Even though it appears that Alloy 2 offers better performance characteristics, Alloy 1 may be less expensive, making it more attractive for certain designs.

Equation 12.1 is a hypothetical equation that would be used to determine the wire diameter for a particular alloy of wire based on the use temperature and current.

$$D = CT_{max}I^n \tag{12.1}$$

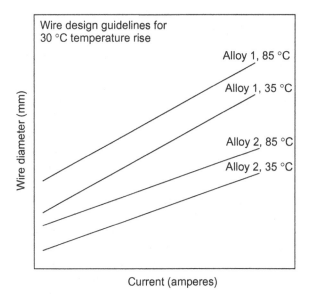

FIGURE 12.1

Graph of hypothetical data for wire diameter versus current.

where D is the wire diameter and C and n are constants that depend on the particular wire material. In addition, the constants may also depend on the microstructure of the material (e.g., grain size or phases present).

Developing a specific design guideline requires that a design team has an understanding about the relationship between the properties of a material and its performance and reliability as used in a product element. In addition, designers must know about the materials' composition, microstructure, surface condition, and defects that correlate to the desired properties. This information can be determined through characterization of the materials combined with verification of the materials to use or product reliability testing.

Design guidelines are developed within companies and by industry standards writing organizations. The benefits of industry standard guidelines are that they are based on the knowledge of a set of people with different experience and they require no investment on the part of the company using the guidelines. The benefit of a company's investment in developing internal design guidelines is that the information is not available to other companies and can provide a competitive advantage.

The benefits of using design guidelines are faster material selection decisions and elimination of the need to conduct materials reliability testing. Both of these help design teams develop their products faster.

12.6 BUDGET FOR MATERIALS ENGINEERING SUPPORT

Companies involved in any aspect of product design, development, or manufacturing should have a budget for materials engineering support. This support includes laboratory support for the analysis of materials and engineering support for all of the situations and decisions discussed in the previous chapters. The amount of support required by any particular company depends on the complexity of the products, the number of new products being developed, the differences in the design requirements from product to product, and the volume of product produced. As any of these items increases, the need for materials engineering support increases.

However, many companies do not have a budget for materials engineering support. Consequently, when decisions or problems arise that require materials engineering support, design, manufacturing, and quality engineers end up struggling to obtain the necessary information rather than spending the money to obtain the required support. At these companies, people accept the "reality" of protracted problems and uninformed decisions rather than obtaining the required materials engineering assistance. The results are missed product deadlines, continued poor quality, and diverted resources.

When problems or decisions that obviously require materials engineering expertise arise, companies without a materials engineering budget for that first look to their engineers or to suppliers for assistance. The engineers typically do not have a materials engineering background and suppliers may or may not have the expertise or time to help properly. Consequently, it takes a long time to make the decision or solve the problem. In both cases, the company and design team spend more time than necessary dealing with the situation, and the outcome is frequently suboptimum.

When trying to solve problems, such as field failures, a company without a materials engineering budget will brainstorm possible root causes. However, without performing an analysis of the failed product and its materials, it is difficult to identify the root cause of the failure.

In contrast, if materials engineering support is included in budgets, then design and manufacturing teams are more likely to seek the necessary assistance in a timely fashion instead of struggling over whether or not to spend the money required to obtain the information necessary to make a decision or resolve a problem.

Materials engineering support can be from internal sources, consultants, and/ or outside analysis groups, including the following forms:

Internal materials engineers. These engineers work for manufacturing companies as design, manufacturing, quality, reliability, and research engineers. In many cases, companies with internal materials engineers also have analysis equipment for evaluating materials features, properties, and performance; surface condition; and defects.

Materials engineering consultants. These engineers can be hired on an as-needed basis.

Materials analysis laboratories. These are service companies that analyze materials properties, composition, microscopic structure, surface condition, and defects. Many of these companies also perform product failure analysis.

12.7 CONSOLIDATE MATERIALS WITHIN AND ACROSS PLATFORMS

When possible, companies should try to use a common set of materials for different products. This applies mainly to products within a common platform. However, it may also apply to products across platforms that have similar mechanical or electrical architectures.

Using a common set of materials offers two benefits. The first is reduced costs associated with buying in volume. The second is being able to use common manufacturing processes within a platform and perhaps across product platforms. This enables manufacturing teams to focus efforts on fewer processes and materials. It also makes it easier to solve manufacturing quality problems and make manufacturing process improvements.

This strategy is similar to not considering the use of all the materials, components, and subassemblies in the world as options when designing a product. However, the idea here is to consolidate the materials for products already in production. Employing this strategy may require some product redesign. In this case, the benefit versus the cost of evaluating and implementing the change will have to be investigated.

CONSOLIDATE MATERIALS WITHIN AND ACROSS PLATFORMS

Index

Printed in the United States
By Bookmasters